CELL BIOLOGY RESEARCH PROGRESS

MEMBRANE POTENTIAL

AN OVERVIEW

MILAN MARUŠIĆ
EDITOR

nova
science publishers
New York

NOTICE TO THE READER

Library of Congress Cataloging-in-Publication Data

ISBN: 978-1-53616-743-6

Published by Nova Science Publishers, Inc. † New York

CELL BIOLOGY RESEARCH PROGRESS

MEMBRANE POTENTIAL

AN OVERVIEW

CELL BIOLOGY RESEARCH PROGRESS

Additional books and e-books in this series can be found
on Nova's website under the Series tab.

CONTENTS

PREFACE

Membrane Potential: An Overview provides trends for the development of novel membranes as separators with applicable properties. The authors offer a comprehensive review of the various types of polymeric materials used as separators either as an electrolyte or not, in different kinds of batteries.

The mathematical formulations of both membrane and fluid parts are reviewed, and general governing equations of the membrane are presented. Also, general formulations of the fluid region and interface conditions between the membrane and fluid sides are presented.

The concluding chapter focuses on the estimation of the plasma membrane potential in yeast by the fluorescence changes of various indicators. The most used of these indicators have been DiSC3(3) and DiSC3(5).

Chapter 1 - Today, energy generation from renewable sources has gained considerable attention to overcome the problems associated with the use of fossil fuels. Energy storage is an essential principle for the efficient use of generated energy by renewable sources. The batteries, as primary stimulating power storage devices, have attracted particular consideration in portable applications such as watches, cell phones, laptops, etc. Improving the performance, achieving a long lifetime, as well as large-scale application capability requires the development and improvement of the components of these systems. In addition to developing new electrode and electrolyte structures, the separator and materials thereof can significantly change the properties of the battery. Despite numerous studies on different types of batteries and their components, researches and publications on the separators are very limited. Properties and structure of the separator, either as an inactive (only in the role of separator) or as an active component (in the function of separator and electrolyte both), play a crucial role in the performance and lifetime of a battery. Separators can be generally classified into porous membranes, modified porous membranes, non-woven fabrics/mat, composite/nanocomposite membranes, and gel-type polymer electrolyte membranes. These structures may be comprised of natural or synthetic polymeric materials as well as minerals, functionalized materials which possess ion transfer or temperature-sensitive/voltage-sensitive properties, and non-functional structures. Besides these, the membrane preparation methods will affect the properties of a separator. The family of polyolefins is the most common group of battery separators that are attracting researchers to study for further modifications ofthese polymers. Preparation of alternative polymers with novel structures, based on natural or synthetic polymers is another goal in this field. The purpose of this chapter is to provide trends in this area for the developing of novel membranes as a separator with applicable properties. It has attempted to offer a comprehensive and general view to the various types of polymeric materials used as separator either as an electrolyte or not, in different kinds of batteries. According to the pioneering researches, the basis of discussions and examples of the chapter is devoted to the separators in Lithium-type batteries, especially Lithium-ion, Lithium-Sulfur, and Vanadium-based Redox flow batteries, especially in the systems that the separators are acting as an electrolyte.

Chapter 2 - Nowadays, the applications of membranes are growing in different areas from modern construction engineering to biological applications. Almost all of these membranes are in contact with different kinds of fluids–stationary or moving–which have a significant effect on the stability of membranes; blood vessels, vehicle airbags, and sails are some notable examples of this subject. Therefore, regarding the importance of the behavior of membranes subjected to fluid in different application, much attention has been drawn to this subject. With this in mind, this chapter presents the basic knowledge of this high-tech area in three main sections. In the first section, initially, the general introduction of this phenomenon is described; then, the potential applications in different fields are introduced. Additionally, a part of this section is dedicated to the dynamical aspects of these problems. The second section deals with the comprehensive mathematical formulations of both membrane and fluid parts. For this purpose, the general governing equations of the membrane are presented; these mathematical equations are given for elastic and hyperelastic membranes. Also, in the next part of this section, both general formulations of the fluid region and interface conditions between the membrane and fluid sides are presented. The third section is devoted to various presented solution methods for coupled fluid-membrane problems and the significant results obtained through analyzing this phenomenon.

Chapter 3 - Estimation of the plasma membrane potential (PMP) in yeast by the fluorescence changes of various indicators have been studied by several groups for many years; with most variable results. The most used of these indicators have been $DiSC_3(3)$ and $DiSC_3(5)$. This contribution explores different dyes studied, methods and incubation media, as well as the different parameters analyzed. Particularly with $DiSC_3(3)$, from rather high to very low values of PMP have been calculated. Recently, the authors reported that fluorescence changes and accumulation of acridine yellow can be used to estimate and obtain actual values of the PMP in yeast. The experiments were performed with an old dye from a flask labeled "acridine yellow" from a commercial source. However, NMR and mass spectrometry revealed that it was thioflavin T. With the pure dye, the experiments were repeated. Also the accumulation of the dye was measured to obtain real values of PMP, mainly based in permeabilizing the cells with chitosan in the absence or presence of an adequate concentration of KCl that allowed to correct the raw data obtained. Hence, more accurate values were obtained. Moreover, results of comparing this dye with others used so far, point to thioflavin T as the best one to follow by fluorescence and measure by its accumulation the plasma membrane potential in yeast.

In: Membrane Potential: An Overview   ISBN: 978-1-53616-743-6
Editor: Milan Marušić   © 2019 Nova Science Publishers, Inc.

Chapter 1

POLYMERIC MEMBRANES
AS BATTERY SEPARATORS

Maryam Mohammadi[1],
Shahram Mehdipour-Ataei[1,]*
and Narges Mohammadi[2]

[1]Faculty of Polymer science, Iran Polymer and Petrochemical Institute,
Tehran, Iran
[2]Faculty of Mechanical engineering, Amirkabir University of
Technology (Tehran Polytechnique), Tehran, Iran

ABSTRACT

Today, energy generation from renewable sources has gained
considerable attention to overcome the problems associated with the use of
fossil fuels. Energy storage is an essential principle for the efficient use of
generated energy by renewable sources. The batteries, as primary
stimulating power storage devices, have attracted particular consideration
in portable applications such as watches, cell phones, laptops, etc.

---
* Corresponding Author's E-mail: s.mehdipour@ippi.ac.ir.

Improving the performance, achieving a long lifetime, as well as large-scale application capability requires the development and improvement of the components of these systems. In addition to developing new electrode and electrolyte structures, the separator and materials thereof can significantly change the properties of the battery. Despite numerous studies on different types of batteries and their components, researches and publications on the separators are very limited. Properties and structure of the separator, either as an inactive (only in the role of separator) or as an active component (in the function of separator and electrolyte both), play a crucial role in the performance and lifetime of a battery. Separators can be generally classified into porous membranes, modified porous membranes, non-woven fabrics/mat, composite/nanocomposite membranes, and gel-type polymer electrolyte membranes. These structures may be comprised of natural or synthetic polymeric materials as well as minerals, functionalized materials which possess ion transfer or temperature-sensitive/voltage-sensitive properties, and non-functional structures. Besides these, the membrane preparation methods will affect the properties of a separator. The family of polyolefins is the most common group of battery separators that are attracting researchers to study for further modifications ofthese polymers. Preparation of alternative polymers with novel structures, based on natural or synthetic polymers is another goal in this field.

The purpose of this chapter is to provide trends in this area for the developing of novel membranes as a separator with applicable properties. It has attempted to offer a comprehensive and general view to the various types of polymeric materials used as separator either as an electrolyte or not, in different kinds of batteries. According to the pioneering researches, the basis of discussions and examples of the chapter is devoted to the separators in Lithium-type batteries, especially Lithium-ion, Lithium-Sulfur, and Vanadium-based Redox flow batteries, especially in the systems that the separators are acting as an electrolyte.

Keywords: battery separator, porous membranes, polyolefin, nanofiber-based membranes, ion exchange membranes, solid electrolyte, gel polymer electrolyte, Li-ion battery, Li-S battery, VRFB

INTRODUCTION

The necessity for the replacement of fossil fuels with clean energy resources has led to much attention toward energy storage systems to reach this goal to practical applications. Batteries using an electrochemical process for energy storage, are one of the most potential candidates for this purpose. The battery was first introduced by Volta in 1800. Now, applications of batteries have revealed in portable devices such as cell phones, watches, walkmans, lap-top computers and notebooks, cameras, toys, as well as stationary and large-scale equipment [1].

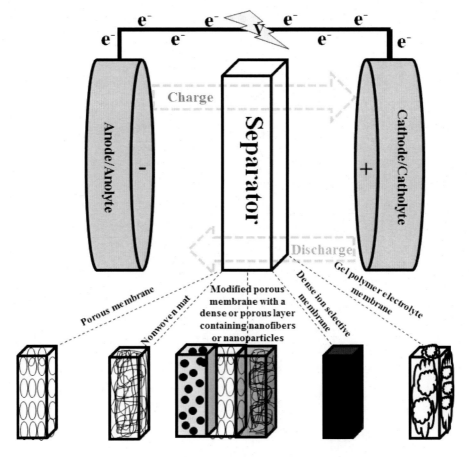

Figure 1. Schematic illustration of typical components of a battery with various types of separators.

Batteries are assembled in different shapes and geometries, so are their components. The key components of a battery are current collector, a positive cathode electrode, a negative anode electrode, and separator, as shown in Figure 1. Despite the development of new batteries and improvement of the structures, there is still no ideal battery with appropriate performance for all operating conditions. In addition to focusing on the development of new electrodes and electrolytes, which are found in many publications, tuning the properties of the separator is a solution to improve the existing technologies and to attain high-performance devices. There is still no single suitable separator for all types of batteries with different geometries and various chemical operating conditions. So understanding the structure-properties relationship of the separator to improve performance and lifetime is one of the future perspectives in this area. However, some review articles are available describing this subject [2-6].

The separator separates the positive and negative electrodes prevents the electrical shortage and allows the ions to transfer by its nature or the electrolyte. Advanced separators are under investigation and production in academic communities and industries around the world. The Asahi Chemical, ATL, Dreamweaver, DuPont, Evonik Litarion, Entek, Freudenberg, Goretex, Hirose, Japan Vilene Company, Kuraray, Leclanché, LG Chemical, Mitsubishi Chemical, Nippon Kodoshi, Panasonic, Polypore, Porous Power, Samsung DI, Senior, SK Energy, Sony, Sumitomo Chemical, Targray, Teijin Aramid, Toray Tonen, Treofan, W-Scope, and Ube Chemical are of the companies which develop and commercialize advanced separators. New separators are either new materials or modified existing separators. The strategies of developing new separators are based on improving their mechanical strength, electrochemical performance, thermal stability, wettability, ion transport, response to overcharge and overheat, barrier properties, easy preparation, and low cost [2, 7, 8]. Proper selection of components and materials is necessary to attain suitable performance.

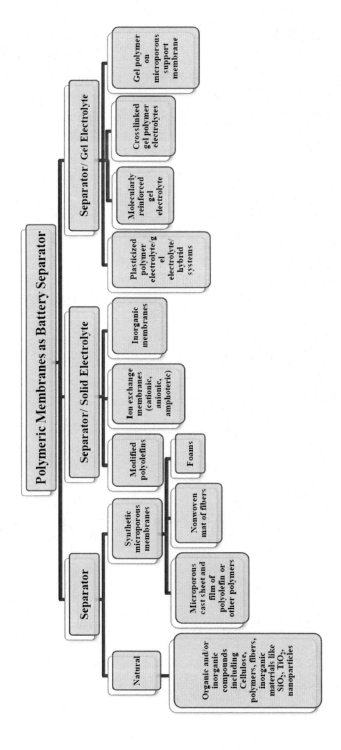

Figure 2. Overall view of membranes classification as a battery separator.

Separators are made up from cellulose [9-11] and cellophane papers to nonwoven mats, foams, ion exchange membranes, and micro/nano-porous flat membranes as single-layer or multilayer comprised of polymer-based or inorganic layers. The overall classification of membranes as battery separators either as a passive or an active component is shown in Figure 2. Regardless of the type of separator, the most commonly used polymers to make separators are polyethylene (PE), polypropylene (PP), polyethylene oxide (PEO), polyacrylonitrile (PAN), poly(methyl methacrylate) (PMMA), polyacrylic acid (PAA), poly(ether imide) (PEI), Nafion, polyvinylidene fluoride (PVDF) and its copolymers including poly(vinylidene fluoride-co-trifluoroethylene), poly(vinylidene fluoride-co-hexafluoropropylene) (PVDF-HFP), and poly(vinylidene fluoride-co-chlorotrifluoroethylene), and gel polymer electrolytes from these polymers. Additionally, blends of these polymers are of interest. Some of the most commonly used blends are PVDF-HFP/PAN, PVDF/PMMA, and PVDF-HFP/PMMA. On the other hand, fillers are divided into active and passive groups. Active fillers are based on lithium salts and participate in the ion transport process. Passive fillers do not participate directly in the transport process. Fillers of this category include inert oxides, ferroelectric materials, clays, carbon materials, and molecular sieves. These materials are compatible with a variety of polymer matrices [2, 12, 13]. Also, eco-friendly materials like poly(lactic acid)/poly(arylene ether ketone) may be used in this application. New separators are categorized as functionalized polyolefins, non-polyolefin polymeric membranes, and mineral types. In the latter, despite the high thermal stability, poor mechanical performance is the existing challenge ahead.

Polymer electrolyte batteries and liquid electrolyte ones have the same chemistry and are expected to have the same power storage capability. However, in practice, the former suffers from more internal resistance that limits the speed of performance. The first successful polymer technology has been developed by Bellcore (now called Telcordia), in which PVDF-HFP has been used as a binder in the active materials as well as the electrolyte. In this system, dibutyl phthalate has been used as a plasticizer to facilitate the manufacturing process, which finally removed from the system. The advantage of these electrolytes is the direct connection of the polymer electrolyte to the surface of the anode and cathode. In this way, the improved output is expected due to the maximized active interface. Therefore, because of the limited studies on battery separators, whether passive or active as separator or separator/electrolyte, this chapter provides a comprehensive and concise overview on the introduction of polymer separators particularly in three types of Lithium-ion (Li-ion), Lithium-Sulfur (Li-S) and Vanadium-based Redox flow batteries (VRFB). To do this, at first, a brief introduction of these three types of batteries and the role of the polymeric membrane is presented. Then, the requirements of the membrane to function in this role are discussed. Subsequently, different types of electrolytes and polymeric membranes in this family will be discussed. Finally, the various types of separators in the role of the anode and cathode separator as an inactive component of the battery, as well as the active member, as a solid polymer electrolyte or gel electrolyte will be explained. The polyolefin separators as the most common family, modified polyolefins, non-olefin membranes, porous membranes, membranes based on nanofibers, and cationic, anionic, and amphoteric ion-exchange membranes, are the constituents of this section. Hope that this chapter provides the information and primary steps which are necessary for beginners who have just started their academic activity in this area as well as expert ones who launch practical studies.

TAKE A LOOK AT THREE TYPES OF LI-ION, LI-S AND VRFB BATTERIES: THE CHALLENGES AND POLYMERIC MEMBRANES

Between different types of batteries and power supplies, Li-ion batteries are highly used for portable applications due to the low price, high energy density, and long life- and storage-time. However, these batteries are not widely available for usage in vehicles due to their high price and low safety. From the materials perspective, all components such as electrode, electrolyte, and separator should be improved to eliminate constraints for non-portable and portable applications. Besides, the thermal runaway event is the safety issue that must be noted. The lithium dendrite penetrates the separator during the cycling and creates a thermal runaway event. In this case, the separator melts at temperatures over 150°C, causing a short circuit inside the battery and releasing the stored energy in the battery. The generated heat in the presence of a liquid electrolyte, which acts as a fuel, causes fire and explosion of the battery. Thermally stable separators, as well as non-flammable solid electrolytes, are needed [1, 14, 15]. The theoretical power density of these batteries is 387 Wh/kg while Li-S batteries have efficient energy of approximately 600 Wh/kg.

In spite of the advantages of Li-S batteries, including the cost-effectiveness due to the usage of sulfur as a petrochemicals byproduct, being eco-friendly, and most importantly, high energy density and high capacity, active sulfur species during electrochemical reactions are converted to soluble polysulfide and solid lithium sulfide/disulfide particles that cause the shuttle effect. Shuttle effect is the most critical problem in these batteries. This effect arises from the migration of dissolved polysulfide anions between the anode and the cathode electrodes and reacting with these electrodes during the charge/discharge processes. So, a dense, thick layer on the cathode will be formed, causing loss of battery performance and capacity, rapid degradation of the electrode components, and consequently, deterioration of the battery. Introducing polysulfide adsorbents in battery components is a solution to limit the penetration of polysulfide species by trapping polysulfide and improving the cyclic stability of these batteries. The shuttle phenomenon is prohibited using novel separators which prevent polysulfide species penetration and migration through physical trapping and/or chemical interaction. Presence of chemical interactions and physical trap simultaneously have a synergistic effect [16].

Redox flow batteries (RFB)s specially applied in cases where high energy efficiency and cost reduction are required. However, the volume and weight of these types of batteries are the dilemmas. The vanadium type has attracted much attention compared to others due to its high cycling, high energy efficiency, high electrical capacity, reasonable price, and chemical stability. VRFBs are suitable sources for large-scale energy storage in the range of kilowatt-hours to megawatt-hours. In this type of battery, the electrolyte contains vanadium solutions in sulfuric acid on both sides of the anode and cathode which uses four oxidation forms of vanadium to transfer electrons through the redox reactions (positive half-cell consists of VO_2^+ which is converted to VO^{2+} upon charging and the negative half-cell contains V^{3+} which is reduced to V^{2+} upon discharging). Therefore, there is no pollution problem in this system, but vanadium permeation from the membrane and from one side to another side of the battery (anolyte and catholyte) may cause spontaneous discharge, low efficiency, and irreversible reduction of capacity. In these batteries, the membrane acts as a separator of positive and negative electrolytes and transfers the charge to complete the circuit. Low cost, chemical stability, being proton-selective against vanadium, sufficient conductivity, and high lifetime are essential characteristics related to the membrane for the commercialization of these batteries. Most studies have been focused on the development of Nafion membranes and proton exchange membranes, while anion exchange membranes and novel types of polymer membranes are desired recently [2, 7, 8, 12, 13, 17-25]. Nafion membranes are widely used in these types of batteries. Nevertheless, despite their advantages, they have limited applications due to the high price and low selectivity against ions,. Porous separators have been attracted much attention as a substitute for ion exchange membranes because of their much lower prices [25-27].

WHAT IS THE ROLE OF POLYMERIC MEMBRANE IN THE BATTERY? HOW DOES IT OPERATE?

The separator is placed between anode and cathode electrodes with opposite polarity or welded together with electrodes in the form of a jellyroll that is called a spiral wound structure. The separator blocks the electrical contact of the electrodes. The same is true for electrolyte reservoir for transportation of ions. The ion conduction is done by the separator, whether through a distinct electrolyte or intrinsic separator nature [2, 7, 8, 17].

Separators are designed to be fitted to four specific batteries. There are some reports for more information on these designs and how to place a separator in them [2].

In Li-ion battery, the first role of the separator is to separate the anode and cathode and provide media for the lithium ion transport during battery operation [8]. In this type of battery, depending on the type of electrolyte, the role of separator varies. In the presence of a liquid electrolyte, separator acts as an inactive component which only separates the anode and the cathode during battery operation. However, its properties determine the energy and power density, cycle life, safety, and therefore, the performance of the battery. On the other hand, in the gel electrolyte, the polymer and the liquid are mixed and the polymer membrane plays both the separator and electrolyte role concurrently. In the former, microporous polyolefin separators are used; whereas, in the latter, separators like PVDF (such as PLION® Cells) are used. For the preparation of this separator, silica and plasticizer are used, then the plasticizer is removed and the porosity is formed. The PVDF-coated polyvinyl pyrrolidone separator is another example of this type of separator. The plasticized electrolyte is another name for this type of separator [7, 28].

In Li-S batteries, the separator is a functionalized porous polymeric membrane and has two roles: first, it separates the anode and cathode parts and prevents short circuit. The second role is to prevent the migration of polysulfide species and battery damage. Conventional polyolefin separators, despite their advantages, are not able to charge and discharge at high temperatures or high voltages, and an internal shortage occurs. Additionally, these membranes are not suitable to prevent permeation of polysulfide species. Therefore, modified functionalized separators are required to block the polysulfide permeation and attain high performance.

In VRFBs as the most common type of RFBs, the separator is an ion conductive and ion-selective membrane, which is placed between two inert electrodes and two solutions containing redox pairs. In this type of battery, side reactions of the electrode are eliminated, and the battery life is theoretically indefinite. The role of the membrane is to separate the positive and negative cells from each other, prevent the mixing of solutions containing vanadium ions, and provide the hydronium ion conduction. Of course, porous membranes can also be used in this battery type provided that porosity content and pore size be adjusted. These types of membranes have low cost, but the optimization of their structures and properties to achieve proper performance and prevent vanadium permeation should be taken into account [29].

REQUIREMENTS OF POLYMER MEMBRANE FOR FUNCTIONING IN BATTERY SYSTEMS

Separator properties determine the performance of the battery. Mechanical and dimensional stability, especially in a spiral wound separator to be capable of wrapping membranes around the electrode at a wide temperature range in the presence of electrolytes are some essential features of an efficient separator. Chemical resistance to electrolyte, impurities, reactants, products and components of the electrode, in addition to thermal stability, ion conduction through the ionic groups present in separator or by immersion in electrolyte, suitable physical strength, appropriate thickness with uniform and defect-free surface to balance the mechanical properties, wettability by electrolyte to reduce the filling time of the cell with the electrolyte and thus able to keep it which results in increasing the battery life in the cases where liquid electrolyte is available, barrier properties for preventing the migration of particles or soluble species between two electrodes, and electrically insulating are another properties required for a membrane to function as battery separator as active and/or inactive component. In general, the properties of the separator should be such to optimize the safety, cost, and performance. In porous membranes, the porosity adjustment is a requirement to prevent the internal shortage, ensuring the suitable permeation of electrolyte and sufficient ion transport. Thin separators with high porosity are used where low power and internal resistance are required; while, thicker separators improve physical strength. On the other hand, one of the ways to increase battery capacity is to reduce the thickness of the separator. Depending on the battery usage, the importance of these general properties varies and depending on the type of battery; the separator should have some specific characteristics that are relevant to the type of battery. For example, in a Li-ion battery, the separator should have a shutdown feature to guarantee safety, and in the alkaline battery, membrane flexibility is required to be welded around the electrode [2, 7, 8, 30].

More specifically, in Li-ion batteries, it should be noted that by increasing the thickness the risk of tearing during cell assembly decreases that is due to increased mechanical strength. On the other hand, the amount of active substance which is used to fill the cell will be lowered. The thin separator occupies less space, giving higher energy and power density and higher surface area, but decreased mechanical strength and safety. The porosity determines the absorption of electrolyte and also ion conductivity. The porosity of membrane in these batteries is about 40%. The pore size must be smaller than the active materials of the electrode and conductive material. The sub-micron pore size is suitable for preventing dendritic lithium penetration. There should be a good interface between the separator and the electrodes to flow the electrolyte properly. Also, in any stages of cell assembly and during the operation of the cell, the separator must not undergo any changes and maintains its integrity. Permeability or McMullin number, which corresponds to the electrolyte and separator resistance respect to the electrolyte resistance, should be in the range of 10-12. The air permeability is expressed by the Gurley number and should be low to achieve good performance. These two numbers represent the tortuosity of the pores. These types of batteries require a high mechanically strengthened separator to provide battery safety at temperatures above 130°C. High thermal stability is also required to achieve more safety and shutdown capability. The shutdown behavior is determined by heating the electrolyte-saturated separator and checking its electrical resistance. The thermal shrinkage of separators for application in Li-ion batteries should not exceed 5% after 60 minutes at 90°C [12, 17, 31]. The minimum mechanical strength and puncture strength is 1000 kg per square meter and 300 g for a 25-micron membrane, respectively. The acceptable value of puncture strength for the prevention of electrode material penetration and the electrical shortage is usually 400 mg/mil. The mix penetration strength, or the separator resistance to particle penetration, is another distinctive feature of the separator that is crucial to prevent the shortage of the battery [7, 8].

Li-S batteries as a subcategory of Li-ion batteries also meet the same general requirements. Besides, in these types of batteries, preventing the permeation of polysulfide species instantaneously with the high conductivity of lithium-ion is required [32, 33].

In addition to the mentioned properties, the polymer membrane in VRFB should also prevent the mixing of the two redox solutions, and prevent the permeation of the vanadium to the opposite side. This effect avoids spontaneous discharging. Also, due to the harsh oxidizing media in these batteries, the membranes should possess high oxidative stability.

Table 1. General properties and battery type related characteristic requirements of the membrane as a battery separator

General Properties	Required Specific Properties Related to the Battery Type
Chemical resistance against electrolyte, impurities, electrodes, etc.	Shut down (Li-ion Battery)
Dimensional stability	Thermal shrinkage < 5% (Li-ion Battery)
Mechanical and physical strength	Polysulfide rejection (Li-S Battery)
Thermal stability	High oxidative stability (VRFB Battery)
Flexibility and easy to handle	Preventing to mix anolyte and catholyte solutions (VRFB Battery)
Porosity content and pore size in porous types	Low vanadium permeability (VRFB Battery)
Ion conductivity (through electrolyte or membrane nature)	
Homogenous, defect-free and optimized thickness	
Barrier property	
Wettability (especially with liquid electrolyte)	
Electrical insulation and preventing battery short-circuit	

In the situations that the polymer separator acts as an electrolyte, in addition to the above requirements, thermal, mechanical and electrochemical stability of the polymer membrane in the temperature range of -20°C to +60°C with a conductivity of 10^{-5} S/cm is necessary.

In Table 1 a brief list of membrane requirements to function in battery either generally or in a specific battery is represented.

POLYOLEFIN TYPE MEMBRANES

Today, the dominant market of separators is occupied by polyolefin membranes [7]. Nonetheless, these separators have low thermal stability, low wettability to the electrolyte, and they are an inappropriate barrier for polysulfide migration in Li-S batteries. PE, PP, and PE/PP blend with a pore size in the micron range are commercially available and are widely used in Li-ion batteries. Asahi, Toray, Entek membrane, Ube, and DSM industries are the leading producers of these membranes. The advantages of these polymers include chemical stability, high porosity, low cost, and excellent processability. However, low thermal stability and a low melting point, which result in permanent degradation caused by thermal runaway, high interface strength, and low wettability by electrolyte are disadvantages. The choice of polyolefin materials in companion with a stretching process to create suitable porosity is one of the main limitations in these types of separators [34]. When using the stretching process to produce porous membranes, creation of shrinkage by polymer chains rearrangement, which can cause battery shortage is unavoidable. These separators have been manufactured via two processing techniques. The first is the wet process, where the granulated PE and wax are melted, mixed, and then extruded. Afterward, it is stretched in two directions, and finally, the wax is washed and removed from the structure. These membranes are used for large cells and have a porosity of about 38-45%, the thickness of 15-32 microns and the weight of 10-20 m^2/g. These membranes have low oxidative stability that may shorten the life of the battery. The second method is the dry process. In this method, PE or PP have been used. Polymer granules melt and extrude. In this method, granules with different crystallinities have been used to control porosity and prepare the pores which are similar to the slit. The membrane is then obtained by the stretching process and annealing the extruded film. In this method, heterogeneity may be seen on the scale of several tens of microns. The membranes comprised of both manners do not have wettability. So, sometimes, a finishing step by corona, plasma, or other treatment techniques is used to modify the membranes and enter the hydrophilic groups [8]. Also, as mentioned above, thermal shrinkage in this type of separators may cause a severe internal shortage, which can be a source of fire, damage, and battery explosion. In this regard, improvements have been made by Celgard, via the production of multi-layer separators with a thermal shutdown mechanism. According to this mechanism, due to the difference in the melting point of PP (165°C) and PE (120-130°C), the PE layer melts up earlier and fills the PP pores, leading to the shutdown of the system that prevents thermal runaway. Although, in practice, it has been observed that the temperature of the battery is higher than both of these temperatures so there is no possibility of this strategy being implemented and the thermal runaway occurs [22]. To improve the oxidative stability of these membranes, a three-layer polyolefin has been developed [31]. The outer layers are from PP, and PE forms the inner layer. PP films have higher oxidation stability than PE; therefore this three-layer membrane has higher oxidative stability and longer lifetime than PE single layer. Separators made up of nanofibers mat are another alternative candidate for these membranes. Some other modified membranes as alternatives for polyolefin membranes are described in the following [7, 35].

SEPARATORS BASED ON MODIFIED POLYOLEFINS

The objective of the study on modified polyolefin separators is to improve their performance by enhancing wettability, thermal stability, porosity and pores integrity, mechanical penetration resistance, safety performance, and facilitating ion conduction without loss of modulus, resistance, and uniformity. The modification of available separators using organic and inorganic materials has some disadvantages. The thickness of the separator increases by adding a coating layer to the separator, which leads to reduced ion conduction, increased resistance of the interface, instability, and delamination of the coating layer, increasing weight and density of the battery, and reducing power density and energy. It should be noted that the non-uniformity is a problem in coated membranes. Thin separators with a thickness of about 6-11 microns, which are provided by a coating of a sol-gel metal oxide layer help to increase the capacity due to low thickness. However, these separators have very high shut-off temperatures that do not meet the Li-ion battery requirement [7]. This family of polymer membranes includes PE coated with mineral nanostructures including glass fibers, metal oxides, silica nanoparticles, polymer/mineral compounds, and other polymers including PVDF, polyvinyl alcohol, polyimide (PI), and others [8, 36]. PE membranes mixed with ceramic, cast films or sprayed layers to provide laminate separators, ceramic-filled mats, ceramic/polymer-coated membranes, nanofiber separators, and Freudenberg separator which is a polyethylene terephthalate mat filled with ceramic are examples of these membranes [37]. Separion® separators combining the polymer and ceramic properties are of commercial products of modified membranes produced by Degussa. This membrane, despite the high cost, shows a similar performance of the polyolefin membranes [17].

Coated polyolefin membranes are usually prepared in two steps: the first step is the production of the porous membrane via wet or dry process, as previously described. In the second step, the coating is coated on one side or both sides. The main problem of this method is the weak adhesion among the coating layer and the membrane. Mechanical interlocking can help to overcome this problem, but another problem is the inadequacy of pores in the interface. Using surfactants, introducing water-soluble functional groups like hydroxyl, carbonyl, carboxyl, amino, imino, sulfonyl, etc. by plasma method or grafting improve the wettability and absorption of the electrolyte. Grafting should be adjusted to optimize mechanical properties and performance. The absorption and maintenance of liquid electrolytes are affected by the amount of grafting and the type of monomer. Grafting of styrene sulfonate or methyl methacrylate monomers on the membrane surface are examples of this modification strategy. Also, to improve the interface between the separator surface and the electrode, a thin polymer layer such as PEO, PVDF and its copolymers [38] can be used to coat the surface of the microporous membranes. This coating transforms into a gel electrolyte by contacting with the liquid electrolyte. It acts as an adherent between the separator and the electrode leading to enhance the absorption and maintenance of the electrolyte. Dipping and spraying methods have been used to apply this coating layer. A heating step has also been used to improve the formation of the gel polymer electrolyte. The drawback of this method is that the polymer layer is dense and prevents the penetration of the liquid electrolyte into the separator pores. To this end, the use of the phase inversion method to form the coating layer instead of the solvent evaporation method, which results in the creation of a microporous polymer layer helps to reduce this problem [17, 28, 39]. PMMA nanoparticles, polyphenols, and polydopamine are also some choices to introduce hydrophilicity. Polydopamine, despite its advantages, including the possibility of applying to various surfaces, several methods of preparation, and the ability to self-polymerized has a high price that limits its use. Tannic acid, with lower price, creates the same level of hydrophilicity and is a good alternative to dopamine. The use of nanofiber coating on the microporous polyolefin membrane improves the adhesion of the membrane and the electrode. Use of PVDF nanofibers coating on a PP microporous membrane is an example of this modification. Polymer blend consisting of PVDF/thermoplastic urethane (TPU) has good electrochemical stability, which is suitable for practical applications. In this blend, TPU is responsible for good dimensional stability [12, 40]. Use of fluorine polymers such as PVDF and its copolymers allows for the adjustment of various microstructures, good mechanical properties, high polarity, and high dielectric constant [38]. The conductivity of polymer-polymer blends depends on the polymer and membrane properties, including porosity, pore size, and its distribution, crystallinity, as well as process conditions such as thermal-induced phase separation, electrospinning, grafting, and the like.

As another modification strategy, we can refer to the usage of aramid nanofibers in improved thermal applications. Polyaniline Nanowire is used for PI separators to improve durability, which also improves the conductivity and performance of the battery. Use of stimuli-response polymer as a coating for promoting safety has been reported. Use of electrochemically active polymers such as poly-4-methoxy triphenylamine preserves the battery from overcharging. Metal oxide nanoparticles, such as alumina, silica, zirconia, titanium are used to improve separator wettability, dimensional stability, thermal stability, and thus battery performance. Between them, zirconia increases the number of micropores and thus, facilitates ion conduction. Metal hydroxides, such as aluminum hydroxide and magnesium hydroxide, are also used to improve the high-temperature performance of separators. These nanoparticles release water at high temperatures, provide greater wettability, higher thermal stability, and higher discharge performance. The choice of coating material and the preparation method of a composite separator should be sufficiently precise to ensure high thermal stability and performance. To this, thermally resistant polymers such as PIs and poly(ether ketone)s have some advantages to control the thermal shrinkage. Polymer/mineral hybrid separators combine the properties of both polymer and mineral, which has a synergistic effect on properties. The structure with silica as core and the poly(ethyl methacrylate) as shell, PI/TiO_2, PI/SiO_2, etc. are some examples of these types of structures. In these structures, the polymer acts as a binder between the mineral nanoparticle and the separator [8, 41]. Also, the functionalization of polyolefin separators that can absorb or convert polysulfide species through physical trapping and chemical interaction is one of the proposed approaches to prevent the shuttle effect in Li-S batteries. Presence of functional groups such as carboxylic acid, or oxygen-rich groups, results in prevention of polysulfide permeation during battery operation. Fiber-glass composite separators apply in Li-S batteries for industrial applications. However, a high thickness of this membrane reduces volumetric density. Metal acids have also been used to prevent the penetration of polysulfide species. Carbon-based materials prevent penetration of polysulfide and activate dead sulfur species due to the conductivity and porosity. These materials also improve performance by lowering battery weight and density [24]. Separator coating with a carbon black layer, inexpensive P carbon black that is commercially available, and Ketjen Black also acts as a current collector at high sulfur content in addition to preventing the polysulfide permeation. To prevent the thickening problem that limits the wettability and lithium-ion conduction, other forms of carbon coatings, including graphene and carbon nanotubes, have been used to modify the separator. Graphene is suitable for hosting sulfur-based materials, and its low weight is an advantage. Due to the complete contact of graphene and sulfur, contact resistance among them also minimizes. Graphene-based materials can also be used for preparing the membrane coating layer. In the presence of graphene oxide, lithium ions are transported by carboxyl groups as hopping sites, and electrostatic interactions reject negatively charged polysulfide species. By reducing graphene oxide and the oxygen-containing functional groups, the conductivity of the membrane and the cathode interface improve and the shuttle of polysulfide is also effectively reduced through the separator [8]. Also, the doping of heteroatoms in carbon coating improves the interaction between polysulfide and carbon coatings and also increases the conductivity of the coating. Carbon nanotube (CNT) with its porous structure, can trap polysulfide species. Due to a large number of micropores and large surface area, single-wall CNT (SWCNT) coatings can prevent the migration of polysulfide and thus improve battery performance. Multi-wall CNT (MWCNT) traps polysulfide species by interacting with sulfur, both chemically and physically. This leads to the formation of additional ion conduction paths by this layer, which improves the properties of the battery. Montmorillonite (MMT) also improves battery performance through electrostatic repulsion with polysulfide in the electrolyte. The combination of carbon materials and polymers is the other type of these separators. Poly (ethylene glycol) (PEG) on the mesoporous carbon-based support on PP is an example of this modification. The benefits of this modified membrane include providing an electrically conductive layer that can act as a current collector by carbon layer and improving the ability to trap polysulfide species physically and chemically due to the presence of PEG. It should be noted that PEG acts as a binder and reduces the thickness of the coating layer [8, 24, 31]. Also, the modification of separators with a layer of ion-selective polymer is another type of these membranes. Simple coating and drying, or casting Nafion solution on the surface of the polyolefin membrane, forms a thin layer of the Nafion membrane as a selective layer that improves the battery performance.

Introduction of an ion-exchange group, a layer by layer assembly of poly (allylamine hydrochloride), and PAA are other examples. Preparation of both ion and size-selective membranes using membranes with a pore size of 0.8 nm compared with 17 nm in the typical Celgard membranes is another way of providing these types of separators. The V_2O_5 with the capability of lithium-ion conduction is suitable for creating an ion-selective layer. In contrast, the use of a thick and dense layer of this material leads to the accumulation of polysulfide species in the cathode. A triple system consists of a macroporous matrix of PP, a thin layer of graphene oxide (GO), and a dense layer of Nafion at a low value of about 0.05 mg/cm^2 which respectively provides the mechanical strength of the system, a coating for matrix, and reduces the migration of polysulfide is another example. In general, the shuttle of polysulfide can be declined with the strategy of creating physical barriers, such as prolonging the permeation through the introduction of tortuosity, similar to the strategy used in a methanol fuel cell to reduce methanol crossover, or by chemical interactions of an ion-selective coating layer [24].

POROUS MEMBRANES

These separators are comprised of a wide range of organic, inorganic, and natural materials with a pore diameter of more than 50-100 °A. Materials include nylon fibers, cotton, polyester, glass, porous polymer films, PE, PP, high-density PE, ultra-high molecular weight PE, poly(tetrafluoro ethylene) (PTFE), poly(vinyl chloride) (PVC), and natural materials, such as rubber, asbestos, and wood [17]. Often, these membranes cannot exchange ions. The common method for the preparation of these membranes is the phase inversion method. The polymer casts on a support and then places in a coagulation bath. Like other applications of porous membranes, the process conditions including coagulation bath temperature, polymeric solution concentration, solvent type, the final thickness of the membrane, etc., have a significant effect on the structure and properties of the membrane. There are some reports for more detailed preparation methods [2, 21, 35, 39].

Porous polymer separators are used in most Li-ion batteries. Despite their advantages, these membranes have low thermal stability, low porosity, and low wettability with electrolyte materials. These disadvantages cause high resistance of the cell and reduce the energy density and cause the low operation speed of these batteries [42].

In Li-S batteries, porous polymer separators provide the requirements, but because of interactions between polysulfide species and the formation of dead sulfur, the battery life shortens. The coating of a thin layer of carbon materials on commercial membranes with diverse methods, such as the slurry coating on the surface, screen printing, and alike is a good way to modify these membranes. Carbon materials have excellent conductivity and high thermal stability. Carbon materials include carbon powder, super P, acetylene black, Ketjen Black, graphene, carbon nanotubes, carbon spheres, carbon nanocomposites, and porous carbon structures. Preparation method of modified membranes with these materials is inexpensive and straightforward, which has attracted much attention in this field. Using different polymer materials with different functionalities and surface properties as a thin layer on the available separators is a way to change surface properties and thus performance. Use of proton exchange polymers, such as Nafion membranes, is a good solution for providing a suitable separator for these types of batteries. This modification, by the introduction of sulfonate groups in the structure, leads to the controledlithium-ion conduction through the mechanism of hopping- besides, the exclusion of negative polysulfide species, leading to the enhanced performance of the battery. The composite of carbon nanotubes with Nafion has been also considered as another solution to prepare the appropriate separator. In this case, the polyolefin membrane is responsible for mechanical strength, carbon material provides conductivity and fills pores of the membrane, and the proton exchange membrane layer prevents the penetration of polysulfide species. Use of mat of Nafion nanofibers also yields another kind of this separator [9, 13].

Besides, a porous separator is the proposed replacement for ion exchange membranes in redox batteries to reduce costs, especially compared to Nafion membranes. In most of these types of batteries, sulfuric acid is used as an electrolyte that provides suitable ion conduction, so it is possible to use porous separators. Although the performance of the battery decreases, it is acceptable against cost reduction. In this case, the hydrolytic stability and ion selectivity of the separator is controlled by tuning the membrane structure [23]. In this type of membrane, the pore size and thickness of the membrane determines the transport velocity of ions with different radii. The proposed mechanisms for vanadium and proton transfer in this type of membrane is the difference in molecular size and radius, density, and proton/vanadium interaction with electrolyte and membrane. Resizing pores using organic/inorganic strategy and increasing the conductivity by an ion-exchange membrane coating is also applicable. Materials used in this field are PAN, poly(ether sulfone), PVDF nanofiltration membranes, polybenzimidazole, $PVDF/SiO_2$, and so on [21, 43]. PVDF ultrafiltration membranes made by the phase inversion method can be suited for this application by changing the solution concentration and other conditions. The smallest obtained pore size is 50 nm. In this membrane, the amount of vanadium permeation is 35% less than Nafion membrane with similar thickness. The pores with hydrophobic walls and porous tortuous pathway are the potential sources of vanadium permeation reduction. In general, the selectivity and proton conduction of this type of membrane are less than that of anion exchange membranes and proton exchange membranes, respectively. Notably, the long-term stability of this type of membrane is the key challenge for application and commercialization. This obstacle is due to the presence of a hydrocarbon backbone which should be exposed to the oxidizing media in these types of batteries. The morphology control of these types of membranes is another issue that must be considered [30, 42].

NANOFIBER-BASED MEMBRANES

Use of nanostructured materials in the area of energy production and storage has attracted much attention in recent years. Fiber-based modified separators allow more electrolytes to be stored, which is attributed to the porosity effect made by nanofibrous morphology. These membranes have higher dimensional stability comparing to coated membranes [7]. Presence of nanofibers with controlled diameter, low density, high surface to volume ratio, and high pore volume are the characteristics of these family that has attracted the attention of studies to this field. The nanofiber structure has better electrochemical properties compared to powder, crystal, nanowire, thin-film, and other structures. A high capacity, stability, and better performance, especially at lower temperatures, are other benefits of these separators.

Nanofibers preparation methods include electrospinning, melt-blowing, bi-component fiber spinning, phase separation, template synthesis, self-assembly, chemical vapor deposition, wet chemical synthesis, free centrifugal or rotary jet spinning nozzle, and force-spinning [31]. Nanofiber membranes made by electrospinning are more porous with smaller pores. Also, the coaxial spinning provides the ability of electrospinning of non-spinnable materials and concentrated solutions in the form of micro and nanofibers [44, 45]. Moreover, the use of dual spinning to prepare composites and polymer blends from these polymers is advantageous [8]. Nanofiber separators have high flexibility and high mechanical stability in which a wide range of polymers such as PI, PAN, PVA, PVDF, polyester, and so on can be used for developing [7]. The review article by Kim et al. [46] may be helpful to gain more information about this structure.

Due to the high porosity and proton pathways formed by the process, these membranes have a high charge/discharge rate in a lithium battery. The major challenge of using electrospinning to produce membrane is the low speed of production [42]. Membranes composed of nanofibers mat show better stability in reducing media than other types of polymers. Nanofiber mat of PVDF, cellulose fibers, ceramic fibers, and polymer nanocomposites thereof are studied. Cellulose fibers have never been commercially available because of the hydrophobic nature of papers and films, their tendency to degrade when exposed to lithium metal, and the ability to form a hole in a thickness of about 100 microns. However, these types of separators are still of interest due to their adjustable morphology and higher thermal stability compared to the polyolefin separators. These types of structures can be used as laminates with low thermally stable materials. In Li-ion batteries, fibers are used as supporting frame [11, 31, 40].

Electrospun fibers have a diameter in the range of several tens of micrometers to several tens of nanometers. The size of pores is predetermined by the fiber diameter, which is a limiting factor. Thus, the fiber diameter is a key parameter determining the transport properties of the mat. Therefore, it is important to understand the relationship between fiber diameters and pore size to reduce dendrite and particles passing. In Li-ion batteries, nanofibers are still not able to compete with microporous films. This is probably due to the inappropriate structure of the pores and the difficulty of producing thin mat with a thickness of lower than 25 microns with acceptable physical properties, good strength, and good uniformity [2]. As mentioned, fiber mat such as cellulose fibers is not widely used in lithium batteries [7]. Composite and multi-layer films of these materials make them suitable for practical applications. For example, a composite of this type of fiber is produced by Asahi Chemical industry [47]. This composite separator consists of cellulose fibers with a diameter of 0.5-5 microns inside a cellulose microporous film with a pore size of 200 microns. These separators have comparable properties to the polyolefin separator [7, 10]. Aramid nanofibers are also suitable for the preparation of separators and composite membranes for Li-ion and RFB batteries. Production of layered composites using aramid fibers provides significant mechanical and thermal properties that result in a longer lifetime. Also, the electrical insulation properties of these materials are unique compared to CNTs, which have comparable dimensions to those of fibers. A solid ion conductive composite based on PEO/aramid nanofibers has been prepared via layer by layer assembly. Formation of hydrogen bonding and subsequent decrease of PEO crystallization is the benefit of this composite, which was a long-term problem for PEO-based electrolytes. Thermal stability up to 450°C, flexibility with high ion conductivity, and dendrite reduction ability are the features of this composite. The aramid nanofibers-based membranes have a permeability of 10 times less than Celgard membrane and a conductivity of 10 times higher than Neosepta membrane. Aramid fibers are also suitable for reducing dendrite and blocking polysulfide species in the Li-S battery

[9]. Preparation of fibers and mat from high-temperature thermally stable polymers such as PTFE, PAN, cellulose, and PVDF is a proposed solution to the thermal runaway. The obstacle of these separators is the large pore size that cannot prevent the penetration of dendrites. For this reason, these structures are used to provide gel polymer membranes that act as an electrolyte as well. So, the ion conductive gel enters the pores of the scaffold, and the dendrite penetration reduces. Ceramic nanoparticles are added to the gel to improve the mechanical strength in the gel areas. In this case, despite the improvement in the mechanical properties of the gel, there is still a problem of dendrite penetration [22]. By adding ceramic nanoparticles to PAN nanofibers and making porous membranes of nanofibers mat, higher absorption of electrolyte, high ion conduction, decreasing the interface resistance of the electrode and electrolyte, and improved chemical stability are achievable [42]. Electrospun membranes made from blend membranes of various polymers, including PAN/PMMA, PAN/PVC, and PVC/PMMA are also applicable. These membranes have high mechanical strength due to the presence of nanofibers [12]. PI nanowire separators have also been manufactured by Jiangxi Advanced Nanofiber Technology Co., Ltd for advanced Li-ion batteries [31].

ADVANCED ION EXCHANGE MEMBRANES

Ion exchange membranes, including proton exchange membranes and anion exchange membranes, are often used in RFB systems. These ion conductive membranes have good mechanical and chemical stability and are proton selective. Proton exchange membranes in comparison with anion exchange membranes have easy synthesis route, adjustable conductivity, and chemical stability. Also, a good understanding of the transport properties of these membranes is attained for practical applications. On the other hand, anion exchange membranes have less vanadium permeability due to the Gibbs-Dunnan effect. Of course, these membranes have less chemical stability and less ion conductivity compared to proton exchange type. This is related to the less mobility of anions comparing to protons. So, lower cycling and voltage efficiency is expected in these membranes.

The redox vanadium media presents a harsh corrosive condition in which it is not possible to use low price conventional membranes. Water transport control is also required to reduce the permeation of vanadium. Use of hydrophobic material in the structure reduces the possibility of interacting with vanadium aqueous solution and increases oxidative stability. In most of the VRFBs, Nafion and Neosepta AHA with high chemical stability and ion conductivity are the common membranes. Nafion membrane transports proton and other monovalent ions well. The problem associated with this membrane is the high price, as well as the high vanadium ions permeability, which causes a rapid drop in capacity with increasing cycles and thus loss of performance. This phenomenon is similar to that of methanol crossover in methanol fuel cells. One of the suggested solutions is to increase the thickness of the Nafion membrane, which reduces the vanadium permeation but decreases the power efficiency due to the increased resistance. However, the first effect is dominant, and the better result is obtained a higher price with a thicker membrane should also be considered [48]. There are also various modifications on Nafion membranes to eliminate or mitigate existing disadvantages; some of them are mentioned below.

In the presence of PTFE with a high hydrophobic nature in the Nafion/PTFE membrane composite, vanadium permeability and excessive swelling decreases, but ion conductivity also drops. To improve the proton conductivity and water absorption, silica particles can be utilized by the sol-gel method. The insertion of a cationic charge layer onto the Nafion membrane is another modification method. The graft copolymerization of dimethylamine aminoethyl methacrylate by the irradiation method on the Nafion membrane is an example of this type of structure. By increasing the graft efficiency, the amount of VO^{2+} ion permeation decreases. Fluorocarbon surfactant brings higher hydrophilicity and thus higher water absorption. Silica nanoparticles and organically modified silicates, PVDF/Nafion composite membrane, multilayer membranes composed of Nafion and sulfonated poly (arylene ether), and Nafion surface modification with polypyrrole, Nafion/graphene oxide composite, sandwich layer, and layer-by-layer structures are examples of other modified membranes have been studied. In all of these membranes, vanadium permeation has been reduced, but, the conductivity of the proton has also been decreased. As the two parameters of vanadium permeation and proton conduction act in the opposite of each other, to obtain a suitable structure, the optimum values should be found. Modification strategies of Nafion are not widely used in practice because of high prices. Use of low-cost, high performance sulfonated poly(arylene ether)s is a suitable alternative in this type of battery [49]. The sulfonated Radel membrane is an example of these membranes. This type of membrane has a lower mechanical, thermal, and chemical stability, excessive swelling, less dimensional stability resulting in a lower lifetime. However, acceptable proton conduction and decreased vanadium permeation are the advantages. The introduction of fluorine groups is a solution to improve the chemical stability of these membranes. In this group, the conductivity of membranes increases with the degree of sulfonation. However, by this increase, the dimensional, mechanical, and oxidation stability decreases. To improve proton conductivity without increasing the degree of sulfonation various strategies, including preparation of phase-separated morphologies on the micro or nano scales, similar to those of proton exchange membrane fuel cells may be effective. Another important parameter is the selectivity of these membranes relative to proton against vanadium. In this regard, it is necessary to design and supply new membranes. Blending of these polymers with other polymers such as PAN, PVDF, PVDF-co-HFP, PIs, and PEIs, as well as preparing block copolymers are examples of alternative membranes. Also, materials such as graphene, graphene oxide, mesoporous silica, zirconium phosphate, sulfonyl phosphonate, titanium, and zinc oxide can be used to produce polymer composites based on sulfonated aromatic hydrocarbons. All of these membranes have lower proton conductivity, but less permeable to vanadium, resulting in gaining higher cycling and higher energy efficiency compared to Nafion. In general, it can be said that all types of membranes used in fuel cells are usable in these batteries. Particularly the basis of the strategies used in methanol fuel cells are applicable to address the proton selectivity over vanadium [20, 22, 23, 50].

On the other side, the three commercial anionic membranes are aminated polysulfone, an ammonium type anion exchange membrane from Asahi Glass Co., and antifouling anion exchange membrane from ASTOM Co. The aminated one has higher resistance with more variation than the other two types. Antifouling membranes have a better selectivity compared to Nafion membranes but show less conductivity. The outcome is the same energy yields. Poly(phthalazinone ether ketone ketone) containing pyridinium groups also shows less vanadium permeation than Nafion 117 by keeping the positive charge via pyridinium groups. Quaternized poly(phthalazinone ether ketone), Cardo poly(ether ketone), polysulfone-based anion exchange membranes with quaternary benzyl trimethyl ammonium groups are other examples of this family. All of these anion exchange membranes also have lower conductivity but higher selectivity in comparison with Nafion. A shorter lifetime than Nafion membrane is expected due to the hydrocarbon backbone in these membranes.

Amphoteric ion-exchange membranes, containing both anion and cation exchange capability are another group of ion exchange separators. These membranes have less but acceptable conductivity, less vanadium permeability, and longer lifetime. PVDF film in which sodium styrene sulfonate and N, N-dimethyl aminoethyl methacrylate have been grafted simultaneously has been applied. Modification of sulfonated poly(fluorenyl ether ketone) by poly(allyl dimethyl ammonium chloride) with positive charge and poly(sodium styrene sulfonate) with a negative charge, and a backbone comprised of brominated fluorinated poly (aryl ether) oxadiazole with 2-aminoethane sulfonic acid side chain as anion and cation exchange groups are examples of these membranes. In these structures, the number of layers affects the properties. It should be noted that the number of studies in this area is limited, and there is still no report on the performance test of this type of membranes in the battery [29].

Use of ion-selective membranes may also be helpful in Li-S batteries. These separators have intrinsic barrier properties against polysulfide species [51]. By using cation exchange membranes, it is possible to prevent the passage of negative polysulfide but allow the lithium-ion to cross. Use of lithiated Nafion is an example of this kind of strategy. Despitethe advantages of Nafion, such as higher cycling stability and shuttle effect reduction, its lithium-ion conductivity is about twice less than that of a liquid electrolyte and is equal to 2.1×10^{-5} S/cm, which results in a loss of efficiency. These properties have been led to the use of Nafion as a selective layer on PP polyolefin membrane and the preparation of a composite membrane [24]. These types of membranes were also introduced in the modified polyolefins section. For example, using a lithiated Nafion ionomer instead of a polyolefin separator in a Li-S battery, presence of sulfonic acid groups, on the one hand, permeates the lithium-ion using the hopping mechanism and on the other hand, prevents the polysulfide shuttling. This type of separator, despite low ion conductivity, results in a good battery performance with high discharge capacity and a suitable number of cycles. The mechanical strength of this type of separators is low compared to polyolefin separators. A thick separator is a proposed solution to this obstacle, as mentioned earlier. To overcome the high cost caused by high thickness, covering a thin layer of Nafion membrane (1-5 microns) on an olefin separator can be performed. This coated separator reduces the shuttle effect with no remarkable change in the conductivity because the coating layer is very thin. So by adjusting the amount of coverage, the proper performance of this type of battery using these types of separators is achievable [8].

THE ELECTROLYTE IN THE BATTERY AND POLYMER MEMBRANE

An electrolyte is referred to as any material containing free ions and is divided into two types of solids and liquids. Generally, the electrolyte contains salts, or dissolved salts in a media. The electrolyte task is to conduct the ions generated in the charge/discharge cycle and to prevent electron conduction. The liquid electrolyte in the battery is a dissolved salt in an ionic or polar liquid that fills the separator [12, 17, 22].

Using a liquid electrolyte, a separator is usually a porous polyolefin such as PP or PE. In these systems, due to the large difference between the polarity of the separator and liquid electrolyte, separator wettability is not suitable by the electrolyte. So, the electrolyte resistance, which determines the performance of the battery, increases [52]. The liquid electrolyte is degraded due to the oxidation reaction in the cathode, and its replacement with a thin polymer membrane can be helpful. If polymer electrolyte membranes are used, both separation and conduction of ions are carried out by the polymer electrolyte.

Solid polymeric electrolytes are divided into two groups: 1) polymer electrolyte, salt-polymer or non-solvent systems 2) hybrid systems, plasticized polymer electrolytes, or gel electrolytes [14, 15, 17, 33, 53]. Polymer electrolytes are alkali metal salts which are present in the main chain or side chain without any liquid. On the other side, polyelectrolytes, are cations or anions covalently bonded to the polymer repeating unit, and these two are not identical. Solid polymer electrolyte systems have a wider operating temperature, greater flexibility for cell design and fabrication, and longer life cycle in Li-S batteries because of the prevention of dissolution of the polysulfide relative to the liquid type [17]. Amorphous regions of polymer structure are the determining factor for ion transport in polymer electrolytes. Hence, ion conduction is controlled by the segmental motions of polymer chain at temperatures above the polymer glass-rubber transition temperature. To achieve good ion conduction in these systems, the amount of salt decomposition must be high. It depends on the total salt concentration in the matrix and usually decreases with increasing salt concentration. The glass-rubber transition temperature increases with increasing salt concentration, and the number of ion carriers comes from chain mobility decreases. So the optimum amount of salt should be found to reach the maximum conductivity. Polymer electrolytes are classified into single types, polymer composites, and polymer blends. The most commonly used fillers for the preparation of polymer composites include inert oxide ceramics such as alumina, titanium, molecular sieve, zeolites, ferroelectric materials, and carbon-based fillers. In the case of blend polymer electrolytes, a polymer helps the electrolyte to function, and the other provides a solid, stable matrix [12]. Linear high molecular weight PEO is an example of polymer electrolyte [15, 54]. The crystallization below the melting point, which leads to poor conductivity, is the shortcoming in this type of polymer. For this purpose, tetra ethylene glycol dimethyl ether with PEO has been used although low capacity has been attained by this system. TiO_2 nanoparticles improve conductivity by reducing the crystallization of the PEO. Unsaturated polyether with tuned structure and properties can eliminate PEO defects by higher ion conductivity in a wider temperature range. Furthermore, copolyethers have an amorphous network that allows the formation of a thin film polymer electrolyte [52]. In addition to linear polymers, polymers in which the PEO is attached to the polymer main chain have been provided. In these systems, the glass transition temperature has been lowered, and more flexible chains have been obtained. As an example, polysiloxanes with PEO side chains are liquid at room temperature with the conductivity of about 10^{-4} S/cm. Although, due to the presence of the Si-O-C side chain, these polymers have low chemical stability. The comb-shaped morphology with the main chain of polyphosphazene with a PEO side-chain has also been studied. In general, these chemicals are complex and also have weak mechanical strength. However, it is possible to attain higher conductivity if one uses either a longer PEO side-chain or a spacer group. In this case, more rigid polymers such as polystyrene or poly(ethyl methacrylate) can be used to improve mechanical properties. Non-solvent systems are not applied in Li-ion and Li-S batteries due to the low ion conductivity. Gel polymer electrolytes or plasticized polymer electrolytes are suitable candidates for these applications [28, 55-57].

Plasticized polymer electrolytes are intermediate between polymer electrolytes and separators. These solid-state polymer electrolytes combine the mechanical properties of a swollen polymer network with high ion conductivity of the liquid electrolyte. Three components of these systems include a polymer, an organic solvent, and a lithium salt. The lithium salt is dissolved in an organic solvent and determines the electrochemical performance and ion conductivity. The polymer matrix provides mechanical strength and preserves the electrolyte solution. Organic solvent swells the polymer matrix, and usually, it forms more than 70% wt. of the electrolyte. It can be said that gel polymer electrolyte contains a trapped liquid in a porous polymer membrane. These electrolytes have a conductivity of about

0.001 S/cm at room temperature. In these systems, the presence of solvent gives plasticization effect and reduces the glass transition temperature, which results in the facilitation of polymer chain motions or in other words, the improvement of the ion conduction. Therefore, the more solvent, the higher ion conductivity is. On the other hand, as excessive solvent weakens the mechanical properties of the polymer, the optimal amount of solvent should be used. It is possible to form a thin film of these electrolytes as well [33]. So, the usage of these systems for Li-ion batteries in applications such as cell phones is possible [52]. Some of the polymers studied in these systems are PEO, PVDF, PMMA, PAN, PVC, and PVDF-HFP. The poor mechanical stability of these polymers is a limiting factor for practical applications. Among all polymers, PVDF-HFP has all the requirements as an electrolyte because it has a high dielectric constant, crystalline vinylidene fluoride units provide mechanical and chemical stability, and the hexafluoropropylene amorphous regions increase the membrane conductivity. Ball milling and phase inversion techniques are used to make these gel polymer electrolytes. The phase inversion method provides the possibility of providing porous membranes with a homogeneous and uniform structure, which bring sufficient electrolyte adsorption and conductivity. In these morphologies interconnected pores to act as a channel for conducting ions are required. Nowadays, it is possible to optimize the membrane preparation conditions or modify the matrix by blending, copolymerization, crosslinking, composite formation, and nanoparticles addition. Crosslinked gel polymer electrolyte and solid polymer electrolyte networks are not willing to crystallize and have good mechanical properties. So, these are suitable for application as thin films and membranes. Several systems have been investigated based on this type of structure, including hydroxyl-terminated polyethers that are crosslinked with di- or tri-isocyanates, allyl-terminated polyethers that are crosslinked with hydrosiloxanes, and polyether triols that are crosslinked with diisocyanates, glycidyl ethers, or UV radiation. As previously explained, the ion conduction in these electrolytes depends on the flexibility of the polymer chain and segmental motions which determine the glass transition temperature. By crosslinking and increasing the crosslink density, the glass transition temperature increases. So, in these systems, less conductivity is expected. The glass transition temperature in these systems is well matched to the Williams-Landel-Ferry (WLF) equation [58]. Of course, in the case of polyphosphazene with PEO side chain, which is allyl-terminated and crosslinked by UV radiation, it is reported that there is no difference between the conductivity before and after radiation. This effect is attributed to the mobility of the polyphosphazene main chain, although crosslinking value has not been reported. Internal lubricant and plasticizer also help to improve the proton conductivity of these systems. Propylene carbonate and ethylene carbonate are among these materials. A large amount of these materials are required for significant improvement, so that they may lead to mechanical weakening and even lack of applicability. However, crosslinked structures with low to moderate crosslink densities, show high swelling, which is an appropriate media for lithium salts. Examples of these structures are oligo(ethylene glycol)-dimethacrylate, which can be swollen with a variety of plasticizers, urethane-crosslinked polyethers, plasticized with N, N-dimethyl formamide and propylene carbonate. Also, using polar comonomers or plasticizers, the polarity of gel increases, and thus, the salt dissolution increases. Using a rigid chain as the backbone and side chains as a liquid matrix for the transport of ionic species, solid polymer electrolyte that is molecularly reinforced with good mechanical strength is obtained. The advantage of using this type of structure in a lithium battery is the ability to produce thin film and thus reduce the resistance of the solid-state lithium battery. These materials are molecular composites, in which both components are mutually compatible at the molecular level. The rigid and flexible chains are aligned together by self-organization. Polyphenylene, as the rigid main chain with flexible side chains of oligo ethylene oxide, is an instance of this architecture. In the case of solid polymer electrolytes without molecular reinforcement, the highest conductivity is attained with the longest side chain, since in this case the PEO matrix crystallinity is prevented. However, the crystallinity of the molecularly reinforced solid polymer electrolytes is disrupted by long side-chain so it should be optimally utilized. The presence of Π bond and the possibility of a reduction reaction that causes electrochemical instability is the disadvantages of these electrolytes [55].

Therefore, the two proposed structures, namely gel polymer electrolytes and molecularly reinforced electrolytes, are the potential solutions for the development of solid-state lithium batteries. The former has high ion conductivity, but poor mechanical properties, and the latter has high stability but low conductivity.

Solvent-free systems without any plasticizer are commonly used for lithium metal batteries. Liquid electrolytes with a separator and plasticized polymer electrolytes are common for Li-ion batteries [17, 28, 52, 59]. To prepare gel polymer electrolyte for Li-ion batteries, usually, a support frame comprised of a nonwoven mat of nanofibers is used. To improve the mechanical strength of polymer electrolytes, microporous polyolefin membranes can also be used as support to provide dimensional stability [60]. Dipping, in situ polymerization, and other conventional filling methods, are used to enter gel polymer electrolytes into the pores. If a copolymer is used, the composition determines the electrolyte adsorption and ion conduction. Given the advantages of gel polymer and microporous membrane, their combination provides better performance against the short-circuit compared with the gel polymer alone. Such a structure is more stable than the microporous membrane as a separator and also a plasticizer as an electrolyte [17, 28, 61].

CONCLUSION AND FUTURE TREND

In this chapter, an overview of the membranes as battery separators was presented. Based on the research results and literature survey, it was concluded that polyolefin membranes are the most popular separators in batteries, especially in Li-ion type. However, progress is followed to prepare a membrane with excellent separation properties together with functioning as an active component. In other words, a membrane which combines the separator and electrolyte properties to a single part is desired. In brief, solid-state batteries are the main focus of studies. To this goal, several types of membranes comprised of modified polyolefins, nanofiber-based membranes, non-olefin modified polymers, ion exchange membranes, solid polymer electrolytes, and gel polymer electrolytes including crosslinked and molecularly reinforced types have been assessed. All of these membranes have their advantages and disadvantages from one battery to the other kind. Nonetheless, the most potential candidate to achieve a suitable product for a wide range of applications is gel polymer electrolytes. Though, studies on different types of custom membranes need to be continued yet, because combining the features of existing membranes with advanced ones may have a synergistic effect.

REFERENCES

[1] Scrosati, Bruno, and Jürgen Garche. 2010. "Lithium Batteries: Status, Prospects and Future." *Journal of Power Sources* 195, no. 9: 2419-30.

[2] Arora, Pankaj, and Zhengming Zhang. 2004. "Battery Separators." *Chemical Reviews* 104, no. 10: 4419-62.

[3] Besenhard, Jürgen O. 2008. *Handbook of Battery Materials*. John Wiley & Sons.

[4] Kinoshita, K, and R Yeo. 1985. *"Survey on Separators for Electrochemical Systems."* LBNL: 18937.

[5] Daniel, Claus, and Jürgen O Besenhard. 2012. *Handbook of Battery Materials*. John Wiley & Sons.

[6] Dell, RM. 2000. "Batteries: Fifty Years of Materials Development." *Solid State Ionics* 134, no. 1-2: 139-58.

[7] Yoshio, Masaki, Ralph J Brodd, and Akiya Kozawa. 2010. *Lithium-Ion Batteries: Science and Technologies*. Springer Science & Business Media.

[8] Xiang, Yinyu, Junsheng Li, Jiaheng Lei, Dan Liu, Zhizhong Xie, Deyu Qu, Ke Li, Tengfei Deng, and Haolin Tang. 2016. "Advanced Separators for Lithium-Ion and Lithium–Sulfur Batteries: A Review of Recent Progress." *ChemSusChem* 9, no. 21: 3023-39.

[9]   Aswathy, R, S Suresh, M Ulaganathan, and P Ragupathy. 2019. "Polysulfide Diffusion Controlled, Non-Shrinkable, Porous, Pan/Pes Electrospun Membrane for High Energy Li-S Battery Application." *Materials Today Energy* 12: 37-45.

[10] Jabbour, Lara, Roberta Bongiovanni, Didier Chaussy, Claudio Gerbaldi, and Davide Beneventi. 2013. "Cellulose-Based Li-Ion Batteries: A Review." *Cellulose* 20, no. 4: 1523-45.

[11] Xu, Quan, Qingshan Kong, Zhihong Liu, Xuejiang Wang, Rongzhan Liu, Jianjun Zhang, Liping Yue, Yulong Duan, and Guanglei Cui. 2013. "Cellulose/Polysulfonamide Composite Membrane as a High Performance Lithium-Ion Battery Separator." *ACS Sustainable Chemistry & Engineering* 2, no. 2: 194-99.

[12] Nunes-Pereira, J, CM Costa, and S Lanceros-Méndez. 2015. "Polymer Composites and Blends for Battery Separators: State of the Art, Challenges and Future Trends." *Journal of Power Sources* 281: 378-98.

[13] Deng, Nanping, Weimin Kang, Yanbo Liu, Jingge Ju, Dayong Wu, Lei Li, Bukhari Samman Hassan, and Bowen Cheng. 2016. "A Review on Separators for Lithiumsulfur Battery: Progress and Prospects." *Journal of Power Sources* 331: 132-55.

[14] Zhang, Wenqiang, Jinhui Nie, Fan Li, Zhong Lin Wang, and Chunwen Sun. 2018. "A Durable and Safe Solid-State Lithium Battery with a Hybrid Electrolyte Membrane." *Nano Energy* 45: 413-19.

[15] Gao, Minghao, Chao Wang, Lin Zhu, Qin Cheng, Xin Xu, Gewen Xu, Yiping Huang, and Junjie Bao. 2019. "Composite Polymer Electrolytes Based on Electrospun Thermoplastic Polyurethane Membrane and Polyethylene Oxide for All-Solid-State Lithium Batteries." *Polymer International* 68, no. 3: 473-80.

[16] Zhang, Sheng S. 2013. "Liquid Electrolyte Lithium/Sulfur Battery: Fundamental Chemistry, Problems, and Solutions." *Journal of Power Sources* 231: 153-62.

[17] Zhang, Sheng Shui. 2007. "A Review on the Separators of Liquid Electrolyte Li-Ion Batteries." *Journal of Power Sources* 164, no. 1: 351-64.

[18] Brodd, Ralph J, Kathryn R Bullock, Randolph A Leising, Richard L Middaugh, John R Miller, and Esther Takeuchi. 2004. "Batteries, 1977 to 2002." *Journal of the Electrochemical Society* 151, no. 3: K1-K11.

[19] Chen, Dongyang, Soowhan Kim, Liyu Li, Gary Yang, and Michael A Hickner. 2012. "Stable Fluorinated Sulfonated Poly (Arylene Ether) Membranes for Vanadium Redox Flow Batteries." *RSC Advances* 2, no. 21: 8087-94.

[20] Schwenzer, Birgit, Jianlu Zhang, Soowhan Kim, Liyu Li, Jun Liu, and Zhenguo Yang. 2011. "Membrane Development for Vanadium Redox Flow Batteries." *ChemSusChem* 4, no. 10: 1388-406.

[21] Wei, Xiaoliang, Bin Li, and Wei Wang. 2015. "Porous Polymeric Composite Separators for Redox Flow Batteries." *Polymer Reviews* 55, no. 2: 247-72.

[22] Tung, Siu On. 2017. *"Aramid Nanofiber Composites for Energy Storage Applications."* Ph.D. diss., University of Michigan.

[23] Perry, Mike L, and Adam Z Weber. 2016. "Advanced Redox-Flow Batteries: A Perspective." *Journal of the Electrochemical Society* 163, no. 1: A5064-A67.

[24] Huang, Jia-Qi, Qiang Zhang, and Fei Wei. 2015. "Multi-Functional Separator/Interlayer System for High-Stable Lithium-Sulfur Batteries: Progress and Prospects." *Energy Storage Materials* 1: 127-45.

[25] Prifti, Helen, Aishwarya Parasuraman, Suminto Winardi, Tuti Mariana Lim, and Maria Skyllas-Kazacos. 2012. "Membranes for Redox Flow Battery Applications." *Membranes* 2, no. 2: 275-306.

[26] Kreuer, Klaus-Dieter. 2013. "Ion Conducting Membranes for Fuel Cells and Other Electrochemical Devices." *Chemistry of Materials* 26, no. 1: 361-80.

[27] Maurya, Sandip, Sung-Hee Shin, Yekyung Kim, and Seung-Hyeon Moon. 2015. "A Review on Recent Developments of Anion Exchange Membranes for Fuel Cells and Redox Flow Batteries." *RSC Advances* 5, no. 47: 37206-30.

[28] Stephan, A Manuel. 2006. "Review on Gel Polymer Electrolytes for Lithium Batteries." *European Polymer Journal* 42, no. 1: 21-42.

[29] Hoang, Tuan KA, and P Chen. 2015. "Recent Development of Polymer Membranes as Separators for All-Vanadium Redox Flow Batteries." *RSC Advances* 5, no. 89: 72805-15.

[30] Venugopal, Ganesh, John Moore, Jason Howard, and Shekhar Pendalwar. 1999. "Characterization of Microporous Separators for Lithium-Ion Batteries." *Journal of Power Sources* 77, no. 1: 34-41.

[31] Agubra, Victor A, Luis Zuniga, David Flores, Jahaziel Villareal, and Mataz Alcoutlabi. 2016. "Composite Nanofibers as Advanced Materials for Li-Ion, Li-O$_2$ and Li-S Batteries." *Electrochimica Acta* 192: 529-50.

[32] Jeddi, Kazem. 2015. *"Polymer Electrolytes for Rechargeable Lithium/Sulfur Batteries."* Ph.D. diss., University of Waterloo.

[33] Zhao, Yan. 2013. *"Polymer Electrolytes for Rechargeable Lithium/Sulfur Batteries."* University of Waterloo.

[34] Ihm, DaeWoo, JaeGeun Noh, and JinYeol Kim. 2002. "Effect of Polymer Blending and Drawing Conditions on Properties of Polyethylene Separator Prepared for Li-Ion Secondary Battery." *Journal of Power Sources* 109, no. 2: 388-93.

[35] Bierenbaum, Harvey S, Robert B Isaacson, Melvin L Druin, and Steven G Plovan. 1974. "Microporous Polymeric Films." *Industrial & Engineering Chemistry Product Research and Development* 13, no. 1: 2-9.

[36] Lee, Hun. *2013. Electrospun Nanofiber-Coated Membrane Separators for Lithium-Ion Batteries.* PhD diss., North Carolina State University.

[37] Hao, Jinglei, Gangtie Lei, Zhaohui Li, Lijun Wu, Qizhen Xiao, and Li Wang. 2013. "A Novel Polyethylene Terephthalate Nonwoven Separator Based on Electrospinning Technique for Lithium Ion Battery." *Journal of Membrane Science* 428: 11-16.

[38] Costa, Carlos M, Maria M Silva, and SJRA Lanceros-Méndez. 2013. "Battery Separators Based on Vinylidene Fluoride (VDF) Polymers and Copolymers for Lithium Ion Battery Applications." *RSC Advances* 3, no. 29: 11404-17.

[39] Huang, Xiaosong. 2012. "A Lithium-Ion Battery Separator Prepared Using a Phase Inversion Process." *Journal of Power Sources* 216: 216-21.

[40] Hwang, Kyungho, Byeongmin Kwon, and Hongsik Byun. 2011. "Preparation of Pvdf Nanofiber Membranes by Electrospinning and Their Use as Secondary Battery Separators." *Journal of Membrane Science* 378, no. 1-2: 111-16.

[41] Jiang, Fengjing, Yu Nie, Lei Yin, Yuan Feng, Qingchun Yu, and Chunyan Zhong. 2016. "Core–Shell-Structured Nanofibrous Membrane as Advanced Separator for Lithium-Ion Batteries." *Journal of Membrane Science* 510: 1-9.

[42] Zhang, Xiangwu, Liwen Ji, Ozan Toprakci, Yinzheng Liang, and Mataz Alcoutlabi. 2011. "Electrospun Nanofiber-Based Anodes, Cathodes, and Separators for Advanced Lithium-Ion Batteries." *Polymer Reviews* 51, no. 3: 239-64.

[43] Zhang, Hongzhang, Huamin Zhang, Xianfeng Li, Zhensheng Mai, and Jianlu Zhang. 2011. "Nanofiltration (NF) Membranes: The Next Generation Separators for All Vanadium Redox Flow Batteries (VRBs)?" *Energy & Environmental Science* 4, no. 5: 1676-79.

[44] Yoon, Jihyun, Ho-Sung Yang, Byoung-Sun Lee, and Woong-Ryeol Yu. 2018. "Recent Progress in Coaxial Electrospinning: New Parameters, Various Structures, and Wide Applications." *Advanced Materials* 30, no. 42: 1704765.

[45] Gong, Wenzheng, Shuya Wei, Shilun Ruan, and Changyu Shen. 2019. "Electrospun Coaxial Ppesk/Pvdf Fibrous Membranes with Thermal Shutdown Property Used for Lithium-Ion Batteries." *Materials Letters* 244: 126-29.

[46] Jung, Ji-Won, Cho-Long Lee, Sunmoon Yu, and Il-Doo Kim. 2016. "Electrospun Nanofibers as a Platform for Advanced Secondary Batteries: A Comprehensive Review." *Journal of Materials Chemistry A* 4, no. 3: 703-50.

[47] Kuribayashi, Isao. 1996. "Characterization of Composite Cellulosic Separators for Rechargeable Lithium-Ion Batteries." *Journal of Power Sources* 63, no. 1: 87-91.

[48] Chen, Dongyang, Michael A Hickner, Ertan Agar, and E Caglan Kumbur. 2013. "Optimizing Membrane Thickness for Vanadium Redox Flow Batteries." *Journal of Membrane Science* 437: 108-13.

[49] Wang, Nanfang, Jingang Yu, Zhi Zhou, Dong Fang, Suqin Liu, and Younian Liu. 2013. "Sppek/Tpa Composite Membrane as a Separator of Vanadium Redox Flow Battery." *Journal of Membrane Science* 437: 114-21.

[50] Mehdipour-Ataei, Shahram, and Mohammadi Maryam. 2019. "Polymer electrolyte membranes for direct methanol Fuel Cells." In *Nanomaterials for Alcohol Fuel Cells* edited by Inamuddin, Tauseef Ahmad Rangreez, Fatih Şen, and Abdullah M. Asiri. 129-158. Materials Research Foundations 49.

[51] Yu, Xingwen, Jorphin Joseph, and Arumugam Manthiram. 2015. "Polymer Lithium–Sulfur Batteries with a Nafion Membrane and an Advanced Sulfur Electrode." *Journal of Materials Chemistry A* 3, no. 30: 15683-91.

[52] Sanchez, Jean-Yves, Fannie Alloin, and Christiane Poinsignon Lepmi. 1998. "Polymeric Materials in Energy Storage and Conversion." *Molecular Crystals and Liquid Crystals Science and Technology. Section A. Molecular Crystals and Liquid Crystals* 324, no. 1: 257-66.

[53] Arya, Anil, and AL Sharma. 2017. "Polymer Electrolytes for Lithium Ion Batteries: A Critical Study." *Ionics* 23, no. 3: 497-540.

[54] Shuo, Zhang. 2044. *Poly (Ethylene Oxide)-Based Composite Polymer Electrolytes for the Secondary Lithium Batteries.* MSC. diss., National University of Singapore.

[55] Meyer, Wolfgang H. 1998. "Polymer Electrolytes for Lithium-Ion Batteries." *Advanced Materials* 10, no. 6: 439-48.

[56] Zhang, Ruisi. 2013. *Advanced Gel Polymer Electrolyte for Lithium-Ion Polymer Batteries.* MSC. diss., Iowa State University.

[57] Liu, Ming, Dong Zhou, Yan-Bing He, Yongzhu Fu, Xianying Qin, Cui Miao, Hongda Du, et al. 2016. "Novel Gel Polymer Electrolyte for High-Performance Lithium–Sulfur Batteries." *Nano Energy* 22: 278-89.

[58] Killis, A, JF LeNest, A Gandini, and H Cheradame. 1981. "Dynamic Mechanical Properties of Crosslinked Polyurethanes Containing Sodium Tetraphenylborate." *Journal of Polymer Science: Polymer Physics Edition* 19, no. 7: 1073-80.

[59] Yu, Ji-Hyun, Jin-Woo Park, Qing Wang, Ho-Suk Ryu, Ki-Won Kim, Jou-Hyeon Ahn, Yongku Kang, Guoxiu Wang, and Hyo-Jun Ahn. 2012. "Electrochemical Properties of All Solid State Li/S Battery." *Materials Research Bulletin* 47, no. 10: 2827-29.

[60] Abraham, KM, M Alamgir, and DK Hoffman. 1995. "Polymer Electrolytes Reinforced by Celgard® Membranes." *Journal of the Electrochemical Society* 142, no. 3: 683-87.

[61] Naderi, Roya. 2016. *Composite Gel Polymer Electrolyte for Lithium Ion Batteries*. MSC. diss., South Dakota State University.

In: Membrane Potential: An Overview   ISBN: 978-1-53616-743-6
Editor: Milan Marušić   © 2019 Nova Science Publishers, Inc.

Chapter 2

APPLICATIONS AND THEORETICAL ASPECTS OF FLUID MEMBRANE INTERACTION

Narges Mohammadi[1], Shahram Mehdipour-Ataei[2],* and Maryam Mohammadi[2]

[1]Faculty of Mechanical Engineering,
Amirkabir University of Technology (Tehran Polytechnique),
Tehran, Iran
[2]Faculty of Polymer Science,
Iran Polymer and Petrochemical Institute, Tehran, Iran

ABSTRACT

Nowadays, the applications of membranes are growing in different areas from modern construction engineering to biological applications. Almost all of these membranes are in contact with different kinds of fluids–stationary or moving–which have a significant effect on the stability of membranes; blood vessels, vehicle airbags, and sails are some notable examples of this subject. Therefore, regarding the importance of the

---

* Corresponding Author's E-mail: s.mehdipour@ippi.ac.ir

behavior of membranes subjected to fluid in different application, much attention has been drawn to this subject. With this in mind, this chapter presents the basic knowledge of this high-tech area in three main sections. In the first section, initially, the general introduction of this phenomenon is described; then, the potential applications in different fields are introduced. Additionally, a part of this section is dedicated to the dynamical aspects of these problems.

The second section deals with the comprehensive mathematical formulations of both membrane and fluid parts. For this purpose, the general governing equations of the membrane are presented; these mathematical equations are given for elastic and hyperelastic membranes. Also, in the next part of this section, both general formulations of the fluid region and interface conditions between the membrane and fluid sides are presented. The third section is devoted to various presented solution methods for coupled fluid-membrane problems and the significant results obtained through analyzing this phenomenon.

Keywords: fluid-membrane interaction, mathematical formulations, solution methods, applications, elastic and hyperelastic membranes

INTRODUCTION

From a mechanical point of view, membranes are lightweight structures that cannot undergo any twisting or bending moments, but only endure tensile stresses. These structures can be imagined as two-dimensional strings, flexible thin plates, or some shell-like configurations subjected to tension. Membranes formed by thin-walled materials must be exerted a pre-stress load to attain a form and to hold a tensile state. Two different ways can be used to apply a pre-stress load and establish a specific form in a membrane. One way is enforcing tension using respective boundary conditions, and another way is through applying internal pressure. This pre-stress load enables membranes to respond and resist against acted forces by adjusting their forms and shapes [1-4].

During the past decades, the applications of membranes in different areas ranging from biological applications to engineering structures have been a growing interest. These thin structures are increasingly recognized as an efficient structure due to considerable merits such as being shapeable and transportable, relatively low cost, and the ease of deployment [1, 5]. Hence, the delicacy and efficiency of these construction elements make them an ideal choice in many practical applications. In most of these applications, membranes are in contact with the stationary or moving fluid that has a significant effect on their behavior that can reflect the requirement to examine fluid membrane interaction problems. To be more specific about these potential applications, extensive practical examples in different disciplines are provided in the following paragraph.

In biomechanics, cardiovascular tissue such as heart, lungs, blood vessels, urinary bladder, valves, as well as, biomedical prostheses such as artificial arteries and organs, and also drug particles like capsules are some cases of membranes subjected to fluid [2, 6]. As regards different fields of engineering, there are also a significant number of membrane structures that are in contact with fluid such as vehicle airbags, water-filled bags, parachutes, sails, tent roofs, windmills, large umbrella, diaphragms in switches and transducers, radio antennas sails, rail cargo, wings, optical reflectors, valveless micropumps, solar arrays, inflatable antennas and reflectors, and the roof of cars [2, 3, 6-10]. All of these notable examples of wide membrane applications in different disciplines are compelling reasons to indicate the importance of the description of the fluid membrane interaction phenomenon.

As stated formerly, most membranes in real-world applications are in contact with fluid; which demands to discern the important issue of stability and oscillation of membranes since the presence of the fluid can dramatically affect membranes behavior and their respective solution method. While a solid structure like membrane is subjected to the fluid, the vibrational frequencies of that decrease compared with when there is no fluid since fluid can act like a damper that can dissipate disturbance. Furthermore, the mass of fluid creates large displacement in the solid part, especially thin ones like membranes. This fact is known as non-dimensional added virtual mass incremental (NAVMI) or added mass effect (AME). In the case of membranes, the low weight and density can exacerbate this effect and may pose some difficulties such as instability and non-convergence of a solution method during numerically solving a problem. This phenomenon broadly occurs in cases that the density of the fluid is close to the structure, or the structure contains internal incompressible fluid. For example, in aeronautical problems, this phenomenon cannot frequently occur; whereas biomechanical problems experience this effect since the blood as the fluid part is considered incompressible fluid and the solid part is a lightweight structure like artery [2, 11-16].

Regarding instability, it may be initiated from a small perturbation in the equilibrium of a system and can be grown by time. For structures subjected flowing fluid, the instability can be categorized as divergence (static instability) and flutter (dynamic instability). As has been mentioned, in the presence of the fluid the frequency of the structure decreases, but, the oscillations transfer to the fluid field can create a velocity potential that causes a significant growth in the total kinetic energy of the coupled system. As the fluid begins to move, the damping of the system, fluid part, dissipates the energy of the structure and subsequently, the frequencies of the structure start to decrease. By exceeding the value of the velocity, this decline continues until the stability occurs within the structure. The point at which the amount of the lowest frequency becomes zero and simultaneously, the sign of damping part changes is considered the onset of the divergence velocity. In this point, the structure sustains a buckling with large amplitude. After divergence instability zone, a further increase in the fluid velocity causes the structure to extract the energy from the fluid part and begin to oscillate with increasingly large amplitude. In other words, the frequency of the structure begins to increase until two frequencies of the structure merge, having the same value, and again, the sign of damping part changes. This point is identified as the onset of the flutter instability that can lead to the self-exciting oscillation of the structure and rapid growth of perturbations [17-25].

It should be pointed out that the instability within the fluid-structure interaction problem must be addressed through nonlinear theories [21]. Additionally, in some FSI problems, the system may not experience divergence instability [26-28]. In other words, with increasing the velocity of the fluid, two modes coalesce, and the vibration of the structure exponentially grows. This state typically occurs in aerodynamic problems.

In both instabilities, as time goes by, besides very large amplitudes, the serious Flow Induced Vibration (FIV) damages can occur that can be catastrophic and disastrous. The FIV possible damages comprise fatigue failure, rupture, collapse, and stress corrosion cracking [29, 30]. This issue can be amplified in a light-weight structure like membranes. Studies regarding the investigation of flutter and divergence instabilities within membrane state the flutter pressure in membrane structures occurs in lower pressures or velocities in comparison with the other structures [31, 32]. This decline is due to lacking flexural stiffness and notable lightness that can deteriorate the effect of the surrounding medium on membranes and cause large distortion [10, 33]. Thus, the issue of stability may heighten the need to precisely and reliably predict the performance and the dynamic behavior of membranes subjected to the fluid to prevent any trouble and ensure performance efficiency. Hence, the concept of instability, as well as the applications of fluid membrane interaction problem offers another convincing reason to deem the fluid membrane interaction phenomenon as a problem of great importance [2, 34, 35].

MATHEMATICAL FORMULATIONS

The utilization of membranes subjected to fluid in different applications entails precisely predicting the behavior of them. For this purpose, two ways of experimental and theoretical methods can be applied to simulate a real problem. Unfortunately, in most cases, a detailed investigation of fluid membrane problems in a frame of experimental work is improbable work. Although experimental results can be a source to validate theoretical responses, they suffer from some drawbacks. The disadvantages of an experimental work include high-costs and difficulties arising from adjusting experiments, being time-consuming, deviation from the realistic situation, and above all, not being capable of explaining the physical and mathematical aspects behind a problem. Therefore, to overcome these limitations, mathematical and physical equations should be employed to enable researchers to predict the dynamic behaviors of these problems expeditiously.

To this aim, in this part, the most general formulations concerning the fluid-membrane interaction problem are presented.

Solid Equations

From a mechanical standpoint, membranes can be deemed as a solid configuration. This implies that the general formulations of solid mechanic also hold for the membranes. Hence, to predict the performance of them, solid equations should be followed.

To describe the behavior of a solid body, three general equations, including Kinematics, Kinetics, and Constitutive equations must be employed. These formulations are for the conditions that there is no source of energy such as heat or electricity. Obviously, in the presence of any source of energy, the corresponding equations must be employed. The following paragraphs provide an overview of the required equations to study the mechanical behavior of the membranes as a solid body [36].

Kinematics Equations

Kinematics equations characterize the geometrical changes without considering the source of these changes. In other words, it states the relationship between strains and displacements. These equations provide a primary step to describe and comprehend the other formulations of a solid configuration. Some of the most common theories to define strain-displacement relationships contain the linear infinitesimal strain tensor, Cauchy-Green deformation tensor, Green-Lagrange strain tensor, and Eulerian Almansi strain tensor [20, 37].

Prior to defining the prevalent representations of strain tensors, herein, some notations concerning the deformation of an element are introduced. Considering a body under external forces moves from the undeformed coordinate X at time $t = 0$, to the deformed coordinate x at time $t > 0$, then, the displacement vector of that body can be expressed as $u_i = x_i - X_i$.

Hereby, the most frequently employed strain tensor representations including right Cauchy-Green deformation tensor C, left Cauchy-Green deformation tensor B, and Green-Lagrange strain tensor E can be defined in terms of the deformation parameters vector for $i, j, k = 1,2,3$ as:

$$C_{ij} = \frac{\partial x_k}{\partial X_i} \frac{\partial x_k}{\partial X_j} \tag{1}$$

$$B_{ij} = \frac{\partial x_i}{\partial X_k} \frac{\partial x_j}{\partial X_k} \tag{2}$$

$$E_{ij} = \frac{1}{2} \left(\frac{\partial x_i}{\partial X_k} \frac{\partial x_j}{\partial X_l} - \delta_{kl} \right) \tag{3}$$

where δ_{kl} is the Kronecker delta. By replacing the displacement vector in Eq. (3), it can be written in the most prevalent format as:

$$E_{ij} = \frac{1}{2} \left(\frac{\partial u_j}{\partial X_k} + \frac{\partial u_k}{\partial X_j} + \frac{\partial u_m}{\partial X_j} \frac{\partial u_m}{\partial X_k} \right) \tag{4}$$

By ignoring the non-linear term, this equation converts to the infinitesimal strain tensor, which may be used for structures with very small displacements. Besides, by introducing the deformation gradient as $F_{i,j} = \frac{\partial x_i}{\partial x_j}$ or $F_{i,j} = \delta_{ij} + \frac{\partial u_i}{\partial x_j}$, they can be rewritten in a compact form as Eq. (5), which can readily represent the relationship between different representations [38, 39].

$$C = F^T F, B = FF^T, E = \frac{1}{2}(FF^T - I) \tag{5}$$

Generally speaking, the main reason for the existing different models is that each of them is appropriate for one type of material model and a specific kind of shape change; For example, among different expressions to describe strains, the right Cauchy-Green deformation tensor offer better formulations, which can be effectively employed for both elastic and hyperelastic materials.

Constitutive Equations

The constitutive equations study the behavior of materials and represent a relationship between two physical quantities. From the solid mechanic point of view, it represents the relationship between stress and strain in a continuum.

Regarding stress tensor, it can be expressed in alternative forms. In infinitesimal deformation like elastic material, all different stress representations are similar, whereas, in large deformation like hyperelastic material, these representations differ from each other. The most frequently used representations of stress encompass the symmetric Cauchy stress tensor, non-symmetric First Piola-Kirchhoff stress tensors, and symmetric second Piola-Kirchhoff stress tensor. Among them, only the Cauchy stress tensor is interpreted as a true stress tensor since it conveys a physical interpretation–the force acted on a unit area. To clarify, Q and q are introduced as the force exerted on the undeformed element and the deformed element, respectively. Additionally, dA and da are also defined as the undeformed and deformed normal element area. Thereby, the Cauchy stress tensor can be expressed based on the deformed configuration (Euler configuration) as $dq = da.\sigma$, whereas, the first and second Piola-Kirchhoff stress tensor are stated based on the undeformed or reference configuration (Lagrangian configuration or material coordinate) as $dq = dA.P$, $dQ = dA.S$ and, respectively.

From a computational standpoint, owing to the symmetry of the second Piola-Kirchhoff stress tensor and a description in reference to the undeformed configuration, it represents a more convenient method to apply and solve in the equilibrium equations. Furthermore, the second Piola-Kirchhoff stress tensor offers a better stress measurement for nonlinear analysis like hyperelastic material, while the Cauchy stress tensor, which is only able to calculate small deformation, is mainly used for elastic material [36-38, 40].

These alternative expressions of stress can be related to each other as below.

$$S = F^{-1}P = jF^{-1}\sigma F^{-T} \tag{6}$$

where j is the relative volume change and defines as $j(X,t) = \frac{dv}{dV} = \det(F)$.

Expressing stress, in terms of strain, greatly depends on the nature of the material. Materials utilized for membranes are of different natures such as elastic, hyperelastic, or viscoelastic materials. Elastic materials only undergo slight deformation, and after removing acted loads, they can attain their former configuration. It implies that the relationship between strain and stress can be presented in a linear format; thus, the well-known generalized Hooke's law is employed to represent the linear stress-strain relationship as below:

$$\sigma = C:\varepsilon + \sigma^0 \tag{7}$$

in which σ and σ^0 are stress tensor and the residual stress tensor, respectively, and ε is the strain tensor. Additionally, the material constant-coefficient C is the fourth-order stiffness tensor of the elastic structures. However, considering the symmetry of the stress and strain tensors besides the concept of the strain energy function, the coefficient matrix reduces to 36 elements with 21 independent variables.

Regarding isotropic elastic materials, since the properties of materials are identical in all directions, the number of independent elastic coefficients reduces to two constants of Lame parameters λ and μ. Hence, the constitutive equation for the stress-strain relationship can be simplified as [36, 41]:

$$\sigma = 2\mu\varepsilon + \lambda tr(\varepsilon)I + \sigma^0 \tag{8}$$

Another category of material is the hyperelastic or non-linear elastic materials. In comparison with linear materials, in hyperelastic materials, the final stress due to a deformation relies on the initial and final states, while in elastic materials; it depends only on the final state. Hyperelastic materials constitute a significant part of membranes' material since most of them comprise polymer-based materials such as soft tissues or rubber materials. These materials undergo large deformation that results in large reversible strains and nonlinear behavior. Therefore, the strain energy function U is required to identify and describe their behavior.

The strain energy function U is expressed based on the principle invariants (I_1, I_2, I_3) and the principle stretch or the square roots (C_1, C_2, C_3) of the eigenvalues of the Cauchy-Green strain tensor.

Although there are many models to express the strain energy function of hyperelastic materials, they can be categorized into three general types. First, the phenomenological model which is appropriate for rubber-like materials and develops a mathematical model based on the experimental data. This model will not be responsive when it is applied to large strains. Some examples concerning this model consist of Mooney Model, Mooney-Rivlin model, Ogden Model, Biderman Model, and Rivlin and Saunders Model. The second type is the micro-mechanical model. This is based on polymer chain entanglements, spatial interactions, and statistical methods. Despite the former model, this is a suitable model to describe large strains. Examples include Veo-hooken Model, the 3-Chain Model, and the 8-Chain Model. The final model is a combination of phenomenological and micro-mechanical models. As a result, this method provides a model that is applicable for both small and large strains as well as it follows experimental data. Examples of this case are the Beda and Chevalier model and the Gent model. Of all models, the Neo-Hooken model and the Mooney-Rivlin model, due to simplicity, are repeatedly employed to model a hyperelastic material, in particular, in fluid membrane problems. The strain energy function of these two models is expressed as:

$$\text{Neo-Hooken model: } U = C_1(I_1 - 3)$$
$$\text{Mooney-Rivlin model: } U = C_1(I_1 - 3) + C_2(I_2 - 3)^2$$

C_1 and C_2 are coefficients which can be attained through experimental results [42].

Concerning elastic material, the strain energy of them can be also obtained through the simple Saint Venant-Kirchhoff model. This model may also be attained through the linear elastic model and Green-Lagrange strain tensor [38, 39, 42].

$$U = \mu E^2 + \frac{\lambda}{2} tr(E)^2 I \tag{9}$$

In the absence of temperature, the second Piola-Kirchhoff stress tensor can be represented in terms of the strain energy function as below:

$$S = \frac{\partial U}{\partial E} = 2\frac{\partial U}{\partial C} = F^{-1}\frac{\partial U}{\partial F} \tag{10}$$

As was mentioned, different stress representations can be related to each other through Eq. (6), therefore, the other representations of stress can be defined with respect to strain energy, likewise.

The last category of materials is viscoelastic materials. This study does not engage with this type of materials. Nonetheless, eager readers can be referred to related subjects [43, 44] to acquire extensive knowledge regarding modeling the equations of this material.

Kinetic Equation

The kinetic or equilibrium equation, which is based on the conservation of momentum, state the equation of motion based on the equilibrium of applied loads. This equation has different forms which all are based on Newton's second law of motion. These equations can be classified into three general approaches: equilibrium, integral, or variational representation. It should be stressed out that all of these different approaches have the same meaning, and employing them in various problems relies on the selected solution method and equations. Three most popular formats of these equations–the conservation of linear momentum, Lagrange equation, and the

Hamilton principle–that are frequently applied in Fluid membrane problems are presented in the following.

The law of the conservation of linear momentum states that a continuous system is in equilibrium if the change of the momentum of that system is equal to the vector sum of all acted forces. It can be mathematically expressed as:

$$\nabla.\sigma + f = \rho \frac{\partial^2 u}{\partial t^2} \tag{11}$$

where f is the body forces, u is the displacement vector, ρ is the density of the solid, and σ is the stress tensor. The conservation of linear momentum, as an equilibrium approach, is a conventional manner to apply in the FSI problem with an elastic solid; nevertheless, this equation has some limitations for more complicated cases [36]. Therefore, for the sake of more suitable equilibrium equations, it would be preferable to search for a more appropriate manner. For this purpose, in the further part, the alternative forms are introduced.

The Lagrangian equation is an integral form to obtain the equilibrium equations of a structure. The original format of Lagrange equations may be expressed as Eq. (12); however, the more general formats of it have been reported in reference [46].

$$\frac{d}{dt}\left(\frac{\partial L}{\partial \dot{q}_n}\right) + \frac{\partial L}{\partial q_n} = Q_n \tag{12}$$

Herein, q and $\dot{q}$ are displacements of the structure and its derivative with respect to the time, and Q is the generalized forces. Additionally, $L = T - \Pi$ is the Lagrange's function in which T and Π are the kinetic energy and the potential elastic deformation energy, respectively. The terms of T and Π may be defined as [45]:

$$\Pi = \iiint_{\Omega} U \, dv \tag{13}$$

$$T = \frac{1}{2}\iint_\Gamma \sum_{i=1}^3 \dot{u}_i \, ds \tag{14}$$

To the best of our knowledge, this method is not hitherto applied in fluid membrane problems. The main reasons to represent this integral approach is mainly due to the application of this approach in some hyperelastic problems. Furthermore, this equation is a base for the Hamilton principle.

The Hamilton principle is a general format of the virtual displacements with respect to time. This principle attempts to attain the minimum of the potential energy for a system; according to this, to establish the equilibrium in a continuous medium, the work that is done by the virtual work fluids–the work of the actual forces moving along virtual displacements–must be zero. Herein, the mathematical formulation of the Hamilton principle, which can apply to all dynamic systems, is expressed as below:

$$\int_{t_1}^{t_2}(\delta T - \delta U + \delta W)\, dt = 0 \tag{15}$$

where δU is the virtual strain energy, δT is the virtual kinetic energy, and δW is the virtual external work; which are defined as follows:

$$\delta U = \iiint_\Omega \sigma : \delta\varepsilon \, dv \tag{16}$$

$$\delta T = \iiint_\Omega \rho \frac{\partial u}{\partial t} \delta\left(\frac{\partial u}{\partial t}\right) dv \tag{17}$$

$$\delta W = \iint_\Gamma f \delta u ds \tag{18}$$

By replacing the terms involved in preceding equations and carrying out some mathematical operation, the constructive equations along with their boundary conditions could be acquired.

In comparison with other methods, the Hamilton principle is of more merits to derive governing equilibrium equations of structures, in particular, when the problem is supposed to solve numerically. This method can successfully be adopted for linear or nonlinear theories which may be suitable for the fluid-membrane problem due to the nonlinearity nature of these problems. Additionally, by this method, the governing equations and boundary conditions can be obtained simultaneously, which can be useful since the boundary conditions associated with constructive equations may be primarily vague. Furthermore, finding approximate solutions, being applicable to the elastic and inelastic materials, presenting mathematical equations based on scalers, and being suitable for all intractable systems are the other advantages of this approach [4, 36, 46].

Membrane Elements

There are two approaches to depict the behavior of a membrane structure as an extremely thin structure. Firstly, one can consider a membrane as a configuration that is not able to tolerate stress couple effect since its tension stiffness is larger than the bending stiffness. In this case, according to the nature of the material and the hypotheses of a membrane element, the above-mentioned equations of a solid configuration–kinematic, kinetic, and constitutive equations–must be employed to each element of the membrane. This approach is able to describe membranes of arbitrary geometries that are made up every type of material. In the second approach, membranes are imagined as thin plates or shell structures that stresses in their thickness direction are ignored. Regarding this case, the conventional equilibrium equations of each geometry, whether shell and thin-plate [36, 47] can be reused without any mathematical manipulations [42, 48]. To clarify, herein, one of these formulations for practicable circular and rectangular membranes along with their respective boundary conditions is given. It should be pointed out that these final formulations all are derived based on the aforementioned formulations [46].

$$T\left[\frac{\partial^2 w(x,y,z)}{\partial x^2} + \frac{\partial^2 w(x,y,z)}{\partial y^2}\right] + F_{ext} = \rho\frac{\partial^2 w(x,y,z)}{\partial t^2} \tag{19}$$

$$T\left[\frac{\partial^2 w(r,\theta,t)}{\partial r^2} + \frac{1}{R}\frac{\partial w(r,\theta,t)}{\partial r} + \frac{1}{R^2}\frac{\partial^2 w(r,\theta,t)}{\partial r^2}\right] + F_{ext} = \rho\frac{\partial^2 w(r,\theta,t)}{\partial t^2} \tag{20}$$

Here, T is the tension of the membrane, w is the transverse displacement, ρ and f are the mass density and external work, respectively, and R is the radius of the circular membrane.

In addition, the boundary conditions of this case can be expressed in three general boundary conditions as follows [46]:

Simply supported: $w = 0$

Free boundary: $\frac{\partial w}{\partial n} = 0$ (n is the normal direction to the membrane surface)

Spring supported: $\frac{\partial w}{\partial n} + kw = 0$ (k is the distributed spring constant)

Fluid Equations

Generally speaking, there are three conservation laws applied to the fluid field in order to determine its properties such as velocity, pressure, and temperature to anticipate either the behavior of the fluid or the effect of the fluid on its ambient. These physics laws comprise the conservation of mass, momentum, and energy. However, owing to the other complexities involved in FSI problems, the vast majority of cases regarding the fluid membrane problems have been conducted based on the two conservations of mass and momentum laws without considering any transformation of energy. For this reason, in the following, these two equations are explained.

The Conservation of Mass

According to the conservation of mass principle, the rate of the change of a mass in a control volume must be equal to the rate of the change of the mass enters and exits the control volume. This principle is often identified as the continuity equation which may be shown in a differential form as:

$$\frac{DM}{Dt} = 0 \tag{21}$$

in which $\frac{D}{Dt}$ is substantial or material derivative that can be expressed for a scaler field φ as:

$$\frac{D\varphi}{Dt} = \frac{\partial\varphi}{\partial t} + u.\nabla\varphi \tag{22}$$

where ∇ is the divergence operator. Therefore, the conservation of mass principle can be written for a control volume as below [49, 50]:

$$\frac{DM}{Dt} = \frac{\partial\rho}{\partial t} + \nabla.(\rho u) = 0 \tag{23}$$

The Conservation of Momentum

In conservation of momentum equation, a similar approach to the conservation of mass may be seen; it means that the change of momentum in a control volume must be equal to the summation of the rate of the momentum enters and exits the control volume, and vector sum of all forces acting on it. It can be said that the conservation of momentum equation is based on Newton's second law of motion for a moving elemental volume of fluid that can be expressed as:

$$\Sigma F = \frac{D(mu)}{Dt} \tag{24}$$

where, F consists of surface forces and body forces. The body forces can appear as gravity, magnetism, Coriolis forces, and electric potential, but, in most of FSI problems, only gravity force ρg is considered as body force acted within a control volume and the other terms are ignored. Additionally, the surface forces contain pressure forces ∇P and viscous forces $\nabla.\tau = \mu\nabla^2 u$. Thus, according to Eq. (22) and Eq. (24) the conservation of mass equation can be rewritten as:

$$\frac{\partial}{\partial t}(\rho u) + \rho u.\nabla u = \rho g - \nabla P + \mu\nabla^2 u \tag{25}$$

Albeit, Eq. (26) is the most general format of conservation of momentum equation, Navier-Stokes equation, in some cases, some characteristics and assumptions of a problem lead more straightforward format of this equation. In what follows, these cases for different types of fluid are mentioned.

Incompressible Fluid

The incompressible fluid flow can be interpreted as the flow in which the change of the density in space and time domain can be ignored. Hence, the continuity equation and conservation of momentum can be rewritten as:

$$\nabla u = 0 \tag{26}$$

$$\rho(\frac{\partial}{\partial t}(u) + u.\nabla u) = -\nabla \vec{P} + \rho g + \mu \nabla^2 u \tag{27}$$

This characteristic of the fluid, compressibility, can be described with the well-known nondimensional Mach number. Mach number is one of the most practical dimensionless numbers, in particular, in aerodynamic problems. This non-dimensional quantity is defined as the ratio of the flow speed to the sound speed as $Ma = \frac{v}{c}$. There are different classifications for different ranges of Mach number [51], but the general classification concerning the compressibility of the fluid is that for incompressible fluid flow $Ma < 0.3$ and, vice versa, for compressible fluid this is $Ma > 0.3$. The incompressible fluid is mainly employed to characterize the blood flow in biological problems.

Inviscid Fluid

In inviscid or frictionless fluid flow, the viscosity can be neglected and considered to be zero ($\mu = 0$). Hence, in Navier-Stokes equations, the viscous term can vanish and the equation can be simplified to the Euler equation as:

$$\frac{\partial}{\partial t}(\rho u) + \rho u.\nabla u = -\nabla \vec{P} + \rho g \tag{28}$$

This format of conservation of mass equation may possess more applications in aerodynamic problems.

Irritational Fluid

The simplest format of the continuity equation is when the fluid flow is steady and irritational as well as inviscid and incompressible. According to the vector analysis, in the irritational fluid flow, the curl, or the cross gradient of the fluid velocity is equal to zero ($\nabla \times u = 0$). So, the velocity of the fluid can be expressed in the format of a scalar function as $u = \nabla \phi$ in which ϕ is named the velocity potential function. This assumption is of great importance since instead of dealing with the vector field, one can more readily solve the equations of the fluid in a scalar field. On the other hand, in the steady fluid, every term is constant over time, so every term concerning time derivative can be eliminated from the equations. By substitution of velocity for velocity potential besides steady assumption, one can arrive at the simplest format of the continuity equation, which called the well-known Laplace equation.

$$\nabla^2 \phi = 0 \tag{29}$$

Likewise, the conservation of momentum equation can be simplified to the Bernoulli equation as:

$$\frac{1}{2}u^2 + \frac{P}{\rho} + gz = 0 \tag{30}$$

in which $u = \nabla \phi$ [50, 51]. Mainly, this type of fluid is not able to represent a realistic situation; nevertheless, it is still employed to describe stationary and steady incompressible fluid flow.

The represented formulations of the fluid are mostly appropriate to study the behavior of the laminar fluid. Since the scope of this study may be too broad, the lengthy subject of turbulent fluid lies beyond the scope of this study.

Coupled Boundary Conditions

In this part, the equations of coupling conditions of an FSI problem are expressed. In order to establish permanent and balanced contact besides full coupling between the common surface of fluid and structure, two kinematic and static boundary conditions at the interface or wet surface must be satisfied simultaneously.

Kinematic coupling condition or continuity of velocity which is known as no-slip or impermeably boundary condition states that the rate of the change of the structure displacement, herein membrane, must be equal with the fluid velocity, Eq. (32). This boundary condition corresponds to the permanent connection of two fields and indicates the equality of the velocities of both fluid and solid.

$$v_i^s = v_i^f \tag{31}$$

For inviscid flow, the equality of velocities only holds in the normal direction to the common surface, so, the Eq. (32) can be modified to:

$$v_i^s.n_s = v_i^f.n_f \tag{32}$$

Additionally, the dynamic boundary condition or continuity of stresses that states the equivalence of forces at the interface can be written as Eq. (34). This condition may also appear in the equilibrium equations of the structure or fluid as an external force induced by the opposite part.

$$\sigma_{ij}^s n_i = \sigma_{ij}^f n_i \tag{33}$$

There is another boundary condition which may not be expressed in some FSI problems; concerning the selected solution procedure, some of the studies do not apply this boundary during the solving process. This is the equality of displacement which expresses the displacement of the structure and fluid should be consistently equal, and any overlap or gap should not happen between two domains [2, 52, 53]

$$x_i^s = x_i^f \tag{34}$$

These coupling boundary conditions are valid provided that the permanent condition always exists and there is no separation or cavitation in the fluid flow.

SOLUTION METHODS

After the comprehension of the governing formulations of fluid membrane interaction problems, it is essential to seek a feasible solution method to predict the behavior of these light-weight structures reliably. To the better introduction of various approaches for fluid membrane interaction problems, we take a more overall look at the different solution methods of FSI problems which can be successfully employed in membrane structures.

There are plenty of computer programs available to predict the behavior of fluids and structures separately–computational fluid dynamic (CFD) and computational structure dynamic (CSD). Nevertheless, the investigation of a coupled FSI problem itself has still remained an ongoing discussion in recent researches [8, 54]. Generally speaking, fluid-structure interaction problems as multidisciplinary problems involve a considerable amount of computational time, mathematical calculations, and modeling efforts. Some difficulties that one may face during the study of an FSI problem include the mutual response of fluid and structure on each other, parabolic-hyperbolic physics of these problems, and the strong nonlinearity nature of these problems especially when the densities of fluid and solid parts are comparable. In the case of fluid-membrane interaction, besides the above-mentioned challenges, there are also some additional intricacies. The high flexibility and small thickness of membranes may cause large amplitudes and strains, which result in highly nonlinear behavior. Furthermore, the hyperelastic behavior of some membranes that leads to nonlinear geometry can be counted as another intricacy. It is said that a thorough and detailed analysis of such a flexible structure can be even more problematic than solving a three-dimensional element [2, 3, 33, 55, 56].

Therefore, in recent decades, many studies are dedicated to this subject to offer an efficient solution method to overcome some challenges involved in the process of solving. Both analytical and numerical methods are developed to adress an FSI problem. Unfortunately, analytical methods are only restricted to some particular cases, while numerical methods cover an extensive range of different methods. Although most numerical methods may be costly and time-consuming, these methods are capable of solving more advanced and complex domains of solids or fluids. The most popular and developed numerical methods to spatially solve include Boundary Element method (BE), Finite Element method (FE), Finite Volume method (FV), Finite Difference (FD), and Spline method. Besides, the Newmark method and the Newton method are frequently employed to solve an FSI problem temporally. With respect to the case study, each of this solution method has its own advantages and disadvantages that should be taken into account.

From the numerical viewpoint, there are various classifications for different numerical approaches that may be adopted for solving a coupled fluid-structure interaction problem; which is somewhat based on different views of the fluid gridding. In accordance with this, the numerical solution methods can be divided into two general classes–monolithic approach and partitioned approach. In the monolithic approach, the entire problem is treated as a single configuration. In other words, the mathematical equations of fluid, solid, and the coupling boundary conditions are expressed as a single system of equations ensuing with a unified algorithm selected to solve them at the same time. This method enjoys advantages, like achieving much better accuracy and stability. Needless to say, solving such a complex system in the format of a unified algorithm not only needs expertise in coding but also involves more time and cost to develop such an algorithm. Besides, it is possible that the solution cannot converge in the case of a considerable added mass effect. Nevertheless, this method can be the right choice for problems of large time-space.

Regarding the partitioned approach, it looks at the fluid and solid fields separately, as two different domains, and subsequently employs different solution methods for each subdomain. The pleasing feature of this approach is the availability of sophisticated codes relating to the fluid and solid domains that reduces the time of developing code and releases applying an extremely specialized algorithm. In other words, instead of dealing with one complicated system, one can readily solve some subsystems. In this approach, information like velocity, displacement, or force is transferred through the interface boundary condition. As a consequence, one can conveniently implement this method for more complex problems and geometries since former sophisticated solvers can be reused. All of these advantages have led to the popularity of this approach, comparing to the former one. In terms of temporal and spatial coupling, the partitioned method can itself be divided into two different groups.

Temporal coupling encompasses two methods of weakly and strongly coupling. In weakly coupling, using a predictor-corrector variable, the fluid and solid domains besides interface boundary conditions are solved once per time step. This results in some benefits and drawbacks. Of downsides of this method are losing numerical stability, in particular, where there is a significant added mass effect. Nonetheless, this method has low computational cost and simplicity to implement.

To hold a strong coupling, at each time step, the fluid and solid domains should be solved until a convergent solution can be attained. Albeit this method provides better accuracy and stability, it suffers from some drawbacks. First, the current fluid and solid solver may undergo some adaptations. Second, when comparing weak coupling, the spatial and temporal solver of fluid and solid have less flexibility. Third, when the added mass effect is considerable, it is possible that this method encounters non-convergent iterative solutions. Fourth, this fashion is a tremendously time-consuming method that demands high storage and CPU time for each iteration. These drawbacks make the strongly coupled method less popular in comparison with the weakly coupled method.

As regards spatial coupling, the partitioned approach can be categorized into two conforming mesh method and the non-conforming mesh method; in terms of the conformity between the body mesh and the interface mesh. In conforming mesh method, as the interface moves, due to the solid deformation, the process of re-meshing is required to follow the behavior of the common surface. As a consequence, it can be a time-consuming and costly procedure to apply in complex geometry since the large deformation of the solid part may happen, and subsequently, a lot of mesh-updating procedures would be needed. On the other hand, remeshing process can produce artificial diffusion and decline accuracy. Nevertheless, this method is of high accuracy and efficiency for turbulent fluid flow. As stated, this method requires a technique to coordinate the fluid and solid solvers; some methods are proposed to follow the moving interface successfully. Examples are the arbitrary Lagrangian-Eulerian formulation (ALE), the space-time finite-element method, and Thin Plate Spline (TPS).

In regard to non-conforming mesh method, the fluid and solid regions are solved independently with their respective solution methods. It means that both fluid and solid fields have their own fixed mesh during the process of solving, and only the interface conditions are applied to their set of equations as constraints in the format of a velocity, stress or force; therefore, there is no need to mesh-updating process to conformity at the interface. As a result, the computational time and cost of the solution procedure may significantly decrease. However, this method loses its stability and accuracy when it is applied to a problem with large deformation.

The non-conforming method can be divided into two other sub-groups, viz., immersed boundary method or immersed body method. The immersed boundary method partially focuses on the fluid domain, whereas the body-fitted method mostly emphasizes the solid domain. In immersed boundary method, the solid is postulated as a thin structure that does not occupy any volume, and the interaction force obtained from the deformation of solid points is acted on the fluid domain as a source term to compute the fluid motion. In contrast, in the immersed domain method, the interface forces due to the artificial fluid are implemented into the Lagrangian configuration of the solid part. This method is mainly efficient for describing bulky and slender structures [11, 22, 52, 55, 57-60].

As noted previously, there exist different geometry and material for membrane as well as various types of fluid. Therefore, it is impossible to achieve a unique solution method for fluid-membrane problems and record the detailed solution procedure; each of case studies requires its own feasible solution method concerning its formulations and conditions. Owing to this fact, the existent various feasible solution methods are outlined in the format of different case studies.

RECENT PRACTICAL RESEARCHES

In this part, the various studies concerning fluid membrane problems are explained. In this respect, the outline of each study containing the formulations, solution methods, and significant results are reported. As there are a lot of applications related to fluid membrane problems that each of them differs from the other ones, it is obviously impracticable to detail all steps of each solution procedure. Therefore, it is attempted to bring the most useful information of each study.

As discussed formerly, the analytical solution for fluid membrane problems as an FSI problem is limited to some simple and specified cases. By considering the simplest format of a fluid consisting irritational, inviscid, incompressible, and steady fluid flow, the analytical solution of membranes with specific shapes, that is, rectangular and circular can be a possible task [61, 62]. Despite the presence of the analytical methods for this sort of membranes, some researchers have also resolved these problems with some numerical techniques such as Finite element [63], and Boundary element [64, 65] methods. The applications of this type of membranes are limited to membranes as transduces and membranes that are backed in fluid-filled tank

Apart from the specific cases above, the rest of the studies of this discipline with more complex geometry and fluid, have been solved numerically. This type covers a wider range of real applications that are mostly based on various gridding methods. In the following, the recent studies concerning this discipline are addressed.

The behavior of closed hyperelastic membranes filled and surrounded with fluid has been done based on the monolithic approach for the internal fluid and the membrane, and partitioned approach for the fluid-filled membrane and the external fluid. It means that the membrane and internal fluid have been discretized based on Lagrangian moving grids, whereas the external fluid is discretized with Eulerian fix grids. The behavior of the fluid part has been described by means of Navier-Stokes equations. For the membrane part, the linear momentum together with Green-Lagrange strain and 2nd Piola-Kirchhoff stress have been employed to study the behavior of the elastic membrane. Both domains have been discretized using Finite Element (FE) method in the space domain and Backward Euler scheme in the time domain. This study has examined some case study, viz., inflation of a square airbag, hanging water bag, underwater balloon, rising balloon, and water-bag offshore reefs for coastal protection [2]. There is a study of membranes contacted with fluid to postulate onset of divergence and flutter instabilities. The membrane has been described by the conventional linear equation of motion for the elastic membranes. These formulations have been given for circular, rectangular, and square membranes in air flow as low subsonic flow. Besides, the inviscid and incompressible fluid flow has been described through 2-D Bernoulli equation and Laplace equation. Boundary element (BE) method and the FE method have been employed to solve the set of equations. The solution method can be deemed as a non-conforming mesh method since the pressure of the fluid has been calculated using boundary element method in terms of the membrane displacement and subsequently has been applied to the membrane elements to estimate the flutter velocity. These membranes can be utilized as a constructive structure like membrane roofs in civil engineer. The results of this study represent the divergence and flutter instabilities of membrane besides, the variation of frequency versus the velocity of the fluid [65]. A similar study as the former one has been done for the 1-D membrane and 2-D fluid both experimentally and theoretically. In a similar fashion, Finite Difference (FD) and BE methods have been used to solve this problem [64]. Another study has aimed to explore the aerodynamic loading on the membranes used as car roofs. By using the simplified format of 1-D equation of flexible panel as the equilibrium equation of membranes, the equation of the membrane has been discretized and solve based on the FD method. The pressure term in membrane equation has been resolved using commercial computational fluid dynamics solver; employing Navier-Stokes (NS) and potential flow (PF) solvers. The beneficial aspects of each solver, like the better accuracy in NS solver and the computational efficiency of the PF solver, besides a comparison of the results of these two methods have been demonstrated in this study. The results reveal the variation of the deformation of membrane versus different streamwise location and direction [66]. The investigation of membrane used as a tent roof has been conducted as a fluid membrane interaction problem. This study has been carried out based on a simulation technique for both structure and turbulent fluid flow. The domain of the structure has been discretized using the FE method with membrane elements. Besides, the Finite Volume method (FV) based on the central difference scheme (CDS) has been selected to discretize domain spatially. The solution procedure is based on the partitioned approach, that is, the distribution of the pressure on the tent roof was calculated based on FV approach and Lagrangian Eulerian technique to advance the location of the wet surface and calculate the displacement of the tent. The solution method reveals that conforming mesh method has been used to solve this coupled problem. Finally, this problem has been solved by means of implicit Newmark-Wilson scheme temporally. The displacements of the tent, owing to the fluid load at different times, have been reported as the final results [8]. The hemisphere membrane inflated with air in a turbulent fluid flow has been simulated based on a conforming mesh method in turbulent flow of wind. The FE solver with Constant Strain Triangle element and FV solver based on the large-eddy technique have been employed to solve membrane element and turbulent flow, respectively. To confirm the fluid and solid grids, the consistent Mortar mapping method has also been applied. In addition, the nonlinear behavior of the silicone membrane has been simulated using the Green-Lagrange strain and St. Venant-Kirchhoff. Finally, the second-order Newmark scheme has been used to solve time depended equations. These membranes can be employed as a tent and petroleum coke bulk storage. Of important results are the frequencies of membrane as well as strain counter at different time instances. It should be noted out that this investigation has been simultaneously carried out experimentally [1, 67]. The partitioned method has been applied to anticipate the membranes as temporary breakwaters filled and surrounded with fluid. To model bottom-mounted shape, the membrane has been modeled as a segment of a circle. The equation of them has been obtained through constitutive law along with Cauchy stress and strain tensor. Fluid equations have been also formulated based on the Laplace's equation and Bernoulli equation. To solve the problem, this study has been developed the BE method for fluid equations and the FE method for membrane equation. The Eulerian-Lagrangian method has been adapted to couple fluid and membrane equations, but the acceleration potential as a predictor-corrector variable has been updated at each time step. Moreover, the fourth-order Runge-Kutta method has been employed to solve the marching problem. This analysis has resulted in calculating the external and the internal pressure distribution on the membrane and demonstrating the deformation and displacement of the membrane besides, comparing the results with a similar experimental study [5, 68]. The analysis of the hyperelastic membrane wing in air vehicle has been scrutinized based on the conforming mesh method. The membrane has been formulated based on the hyperelastic Mooney-Rivlin model, Green-Lagrange strain tensor, second Piola-Kirchhoff stress tensor, and virtual work principle. This study has been carried out using the FV method based on the pressure-based solver (PISO) for the turbulent fluid and also triangular-shaped finite elements for the membrane. Furthermore, to update the moving grids, Thin Plate Spline (TPS) fashion has been adopted. Finally, both explicit central finite difference and implicit Newmark algorithm have been employed to solve this problem temporally. The result of this study illustrates the dynamic response of the wing membrane due to the aerodynamic pressure at different time steps [48, 54]. It is noteworthy that the experimental work on membrane airfoil has been also conducted in another similar study [69]. Another investigation regarding the fluid membrane interaction is on elastic membranes that are applicable as photographic film or plastic tape contacted with the viscous fluid. In accordance with conforming mesh method, Space-time (DSD/ST) finite element method has been used to solve the fluid membrane interaction problem. Results include the mechanical behavior of the membrane such as strain, velocity, displacement at different time step [70]. A study to address the biological membrane, arteries containing blood flow, in contact with fluid has been carried out based on the conforming mesh method approach. The hyperelastic membrane has been formulated with the Neo-Hookean material model. Besides, the membrane's equation has been derived based on the virtual work principle, along with Piola-Kirchoff stress tensor and Cauchy-Green strain tensor. To solve the boundary volume problem, two discretization methods, including the FV method and the FE method, have been adopted to solve Navier-Stokes equations and membrane dynamic equation, respectively. Furthermore, the arbitrary Lagrangian-Eulerian formulation has been used to exchange information between the fluid and solid models. Finally, the implicit Newmark time integration scheme has been selected to solve the initial value problem [71]. There is an investigation concerning hyperelastic membranes subjected to flow that can be useful to examine membrane capsules and red blood cells. This is based on the implicit immersed boundary method. The equation of the elastic membrane has been described through the linear constitutive equation along with Cauchy-Green strain tensor, and Neo-Hookean model. In order to discretize the domain of membrane and fluid, the FE method and the FV method have been applied, respectively. Moreover, to capture the interaction between membrane and fluid, the cost-efficient method of Jacobian-free Newton-Krylov (JFNK) has been introduced. The significant results include the demonstration of the deformation of red blood cells and membrane spheroidal capsules [72]. The results of this study have been also compared with previous experimental work [43]. A study in order to examine the blood flow in arteries has been conducted using the partitioned approach. This study has been described the arteries as a cylindrical shell with considering the small thickness and neglecting elastic bending terms, and the blood as the 2-D fluid. Both equations have been discretized by means of the FE method. In addition, the implicit Euler and mid-point schemes have been selected to solve the fluid and membrane problem temporally, respectively. To effectively solve the problem, this study has been introduced an algebraic fractional scheme to enhance convergence and numerical stability. The deformation of arteries at different time step has been provided as the final results [73]. The behavior of membrane subjected to the fluid has been examined for some practical examples such as flapping flags, and the radial fluid-inflated cylindrical membrane. The FE method, together with the ALE method, has been implemented to discretize and connect fluid and solid equations. Besides, the Newton-Raphson method has been applied to solve the time-dependence problem. Results of this study illustrate the deformation of each membrane at different time steps [74].

CONCLUSION

This chapter aimed to provide an overview of fluid membrane interaction problems, with particular emphasis on the membrane side. This outline was given in the format of different sections, including the introduction, formulation, solution method, and the review part. Accordingly, the fluid membrane interaction as a multidisciplinary problem is of crucial importance in different points of view. Firstly, membranes contacted with fluid have diverse and extensive applications in different real-world areas such as biomedical prosthesis, inflatable structure, and the roof of cars. Secondly, the significant effect of fluid mass on membranes, as light-weight structures, may become problematic, in particular, in the process of solution. Lastly, the important issue of instability–divergence and flutter instability–may lead to serious damages in membrane structure.

Further to this, in the formulation part, the general equation of the membrane as a solid structure and fluid was mentioned practically; this enables researchers to apply these formulations appropriately in extensive ranges.

As regards the solution method part, an overview of possible solution approaches for a coupling problem, such as partitioned and monolithic methods along with temporal and spatial coupling methods relating to the former approach was provided. It can be inferred that the selected solution approach highly depends on the complexity of the membranes, nature of the fluid, the importance of the accuracy of the problem, and the crucial factor of the computational cost.

Finally, according to the review of recent advances of this discipline and the best of the authors' knowledge, this review reveals the ample room for more potential applications regarding this growing subject. Some examples include: simulating more challenging problems such as the study of the interaction of porous membranes and fluids, the impacts of shock-wave on membranes, the interaction of non-Newtonian fluid and membranes, or the interaction of viscoelastic membranes and fluids. Moreover, with respect to the weak points of each proposed solution method, the current solution methods can be improved, and more innovative solutions can be presented.

REFERENCES

[1] De Nayer, G., Apostolatos, A., Wood, J. N., Bletzinger, K.-U., Wüchner, R. & Breuer, M. (2018). Numerical studies on the instantaneous fluid–structure interaction of an air-inflated flexible membrane in turbulent flow. *Journal of Fluids and Structures, 82,* 577-609.

[2] Ryzhakov, P. B. & Oñate, E. (2017). A finite element model for fluid–structure interaction problems involving closed membranes, internal and external fluids. *Computer Methods in Applied Mechanics and Engineering, 326,* 422-445.

[3] Vázquez, V. & Gerardo, J. (2007). *Nonlinear Analysis of Orthotropic Membrane and Shell Structures Including Fluid-Structure Interaction.* PhD diss., Technical University of Catalonia.

[4] Rao, S. S. (2007). *Vibration of continuous systems* (Vol. *464*): Wiley Online Library.

[5] Liu, C. & Huang, Z. (2019). A mixed Eulerian-Lagrangian simulation of nonlinear wave interaction with a fluid-filled membrane breakwater. *Ocean Engineering, 178,* 423-434.

[6]   Muha, B. & Čanić, S. (2015). Fluid-structure interaction between an incompressible, viscous 3D fluid and an elastic shell with nonlinear Koiter membrane energy. *Interfaces and free boundaries*, *17*(4), 465-495.

[7]   Gascón-Pérez, M. (2017). Acoustic influence on the vibration of a cylindrical membrane drum filled with a compressible fluid. *International Journal of Applied Mechanics*, *9*(05), 1750072.

[8]   Glück, M., Breuer, M., Durst, F., Halfmann, A. & Rank, E. (2001). Computation of fluid–structure interaction on lightweight structures. *Journal of Wind Engineering and Industrial Aerodynamics*, *89*(14-15), 1351-1368.

[9]   Jenkins, C. H. & Leonard, J. W. (1991). Nonlinear dynamic response of membranes: state of the art. *Applied Mechanics Reviews*, *44*(7), 319-328.

[10]   Gascón-Pérez, M. (2018). Interactions of an oscillating rectangular membrane with a compressible fluid. *International Journal of Applied Mechanics*, *10*(02), 1850016.

[11]   Idelsohn, S. R., Del Pin, F., Rossi, R. & Oñate, E. (2009). Fluid–structure interaction problems with strong added-mass effect. *International journal for numerical methods in engineering*, *80*(10), 1261-1294.

[12]   Desmars, N., Tchoufag, J., Younesian, D. & Alam, M. R. (2018). Interaction of surface waves with an actuated submerged flexible plate: Optimization for wave energy extraction. *Journal of Fluids and Structures*, *81*, 673-692.

[13]   Kwak, M. K. (1996). Hydroelastic vibration of rectangular plates. *Journal of Applied Mechanics*, *63*(1), 110-115.

[14]   Kwak, M. (1991). Vibration of circular plates in contact with water. *Journal of Applied Mechanics*, *58*(2), 480-483.

[15]   Yildizdag, M. E., Ardic, I. T., Demirtas, M. & Ergin, A. (2019). Hydroelastic vibration analysis of plates partially submerged in fluid with an isogeometric FE-BE approach. *Ocean Engineering*, *172*, 316-329.

[16] Van Brummelen, E. (2009). Added mass effects of compressible and incompressible flows in fluid-structure interaction. *Journal of Applied mechanics*, *76*(2), 021206.

[17] Shankar, V. (2015). Stability of fluid flow through deformable tubes and channels: an overview. *Sadhana*, *40*(3), 925-943.

[18] Bahaadini, R. & Hosseini, M. (2016). Nonlocal divergence and flutter instability analysis of embedded fluid-conveying carbon nanotube under magnetic field. *Microfluidics and Nanofluidics*, *20*(7), 108.

[19] Amabili, M., Pellicano, F. & Païdoussis, M. (1999). Non-linear dynamics and stability of circular cylindrical shells containing flowing fluid. Part I: stability. *Journal of sound and Vibration*, *225*(4), 655-699.

[20] Amabili, M. (2008). *Nonlinear vibrations and stability of shells and plates*: Cambridge University Press.

[21] Bochkarev, S. & Matveenko, V. (2011). Natural vibrations and stability of shells of revolution interacting with an internal fluid flow. *Journal of Sound and Vibration*, *330*(13), 3084-3101.

[22] Hou, G., Wang, J. & Layton, A. (2012). Numerical methods for fluid-structure interaction—a review. *Communications in Computational Physics*, *12*(2), 337-377.

[23] Zuo, Q. & Schreyer, H. (1996). Flutter and divergence instability of nonconservative beams and plates. *International Journal of Solids and Structures*, *33*(9), 1355-1367.

[24] Kwak, M. & Kim, K. (1991). Axisymmetric vibration of circular plates in contact with fluid. *Journal of Sound and Vibration*, *146*(3), 381-389.

[25] Fernández, M. A. & Le Tallec, P. (2003). Linear stability analysis in fluid–structure interaction with transpiration. Part II: Numerical analysis and applications. *Computer methods in applied mechanics and engineering*, *192*(43), 4837-4873.

[26] Zhang, L., Song, Z. & Liew, K. (2016). Computation of aerothermoelastic properties and active flutter control of CNT reinforced functionally graded composite panels in supersonic airflow.

Computer Methods in Applied Mechanics and Engineering, *300*, 427-441.

[27] Attar, P. J., Gordnier, R. E., Johnston, J. W., Romberg, W. A. & Parthasarathy, R. N. (2011). Aeroelastic analysis of membrane microair vehicles—part I: flutter and limit cycle analysis for fixed-wing configurations. *Journal of Vibration and Acoustics*, *133*(2), 021008.

[28] Dowell, E. H. (1967). Nonlinear oscillations of a fluttering plate. II. *AIAA journal*, *5*(10), 1856-1862.

[29] Mohammadi, N., Asadi, H. & Aghdam, M. M. (2019). An efficient solver for fully coupled solution of interaction between incompressible fluid flow and nanocomposite truncated conical shells. *Computer Methods in Applied Mechanics and Engineering*, *351*, 478-500.

[30] Ter Hofstede, E. (2015). *Numerical Study of Fluid Structure Interaction in Nuclear Reactor Applications*. MSc diss., Delft University of Technology.

[31] Anderson, W., Messiter, A. F. & Spriggs, J. H. (1969). Membrane flutter paradox-An explanation by singular-perturbation methods. *AIAA Journal*, *7*(9), 1704-1709.

[32] Haruo, K. (1975). Flutter of hanging roofs and curved membrane roofs. *International Journal of Solids and Structures*, *11*(4), 477-492.

[33] Soares, R. M. & Gonçalves, P. B. (2018). Nonlinear vibrations of a rectangular hyperelastic membrane resting on a nonlinear elastic foundation. *Meccanica*, *53*(4-5), 937-955.

[34] Jeong, K. H., Yoo, G. H. & Lee, S. C. (2004). Hydroelastic vibration of two identical rectangular plates. *Journal of Sound and Vibration*, *272*(3-5), 539-555.

[35] Kerboua, Y., Lakis, A., Thomas, M. & Marcouiller, L. (2008). Vibration analysis of rectangular plates coupled with fluid. *Applied Mathematical Modelling*, *32*(12), 2570-2586.

[36] Reddy, J. N. (2004). *Mechanics of laminated composite plates and shells: theory and analysis*: CRC press.

[37] Helnwein, P. (2001). Some remarks on the compressed matrix representation of symmetric second-order and fourth-order tensors.

Computer Methods in Applied Mechanics and Engineering, *190*(22-23), 2753-2770.

[38] García González, D. (2016). *A continuum mechanics framework for hyperelastic materials: connecting experiments and modelling.* PhD diss., Universidad Carlos III de Madrid.

[39] Guo, Z. (2006). *Computational modelling of rubber-like materials under monotonic and cyclic loading.* PhD diss., Delft University of Technology.

[40] Taber, L. A. (2004). *Nonlinear theory of elasticity: applications in biomechanics*: World Scientific.

[41] Sadd, M. H. (2009). *Elasticity: theory, applications, and numerics*: Academic Press.

[42] Patil, A. (2016). *Inflation and Instabilities of Hyperelastic Membranes.* PhD. Diss, KTH Royal Institute of Technology.

[43] Christensen, R. (2012). *Theory of viscoelasticity: an introduction*: Elsevier.

[44] Brinson, H. F. & Brinson, L. C. (2008). *Polymer Engineering Science and Viscoelasticity*: Springer.

[45] Breslavsky, I. D., Amabili, M. & Legrand, M. (2014). Nonlinear vibrations of thin hyperelastic plates. *Journal of Sound and Vibration*, *333*(19), 4668-4681.

[46] Meirovitch, L. (1997). *Principles and techniques of vibrations*, (Vol. *1*), Prentice Hall New Jersey.

[47] Timoshenko, S. P. & Woinowsky-Krieger, S. (1959). *Theory of plates and shells*: McGraw-hill.

[48] Lian, Y., Shyy, W., Ifju, P. & Verron, E. (2002). A computational model for coupled membrane-fluid dynamics. Paper presented at the *32nd AIAA Fluid Dynamics Conference and Exhibit*, p. 2972.

[49] Versteeg, H. K. & Weeratunge M. (2007). *An introduction to computational fluid dynamics: the finite volume method.* Pearson education.

[50] Fox, R. W., McDonald, A. T. & Pitchard, P. (2006). *Introduction to Fluid Mechanics*, 2004. In: John Wiley & Sons, Inc.

[51] White, F. M. (2011). *Fluid Mechanics*: McGraw Hill.

[52] Hosters, N., Helmig, J., Stavrev, A., Behr, M. & Elgeti, S. (2018). Fluid–structure interaction with NURBS-based coupling. *Computer Methods in Applied Mechanics and Engineering*, *332*, 520-539.

[53] Gatzhammer, B. (2014). *Efficient and flexible partitioned simulation of fluid-structure interactions.* PhD diss., Technische Universität München.

[54] Lian, Y. & Shyy, W. (2003). *Three-dimensional fluid-structure interactions of a membrane wing for micro air vehicle applications.* Paper presented at the 44th AIAA/ASME/ASCE/AHS/ASC Structures, Structural Dynamics, and Materials Conference.

[55] Gilmanov, A., Le, T. B. & Sotiropoulos, F. (2015). A numerical approach for simulating fluid structure interaction of flexible thin shells undergoing arbitrarily large deformations in complex domains. *Journal of Computational Physics*, *300*, 814-843.

[56] Oden, J. & Sato, T. (1967). Finite strains and displacements of elastic membranes by the finite element method. *International Journal of Solids and Structures*, *3*(4), 471-488.

[57] Kim, W. & Choi, H. (2019). Immersed boundary methods for fluid-structure interaction: A review. *International Journal of Heat and Fluid Flow*.

[58] Van Loon, R., Anderson, P., Van de Vosse, F. & Sherwin, S. (2007). Comparison of various fluid–structure interaction methods for deformable bodies. *Computers and Structures*, *85*(11-14), 833-843.

[59] Matthies, H. G. & Steindorf, J. (2003). Partitioned strong coupling algorithms for fluid–structure interaction. *Computers and structures*, *81*(8-11), 805-812.

[60] Nicoud, F., Mendez, S., Deplano, V., Gerbeau, J. F., Segers, P. & Moureau, V. (2016). *Fluid-structure interaction problems involving deformable membranes: application to blood ows at macroscopic and microscopic scales.* PhD diss., Ghent University.

[61] Tariverdilo, S., Mirzapour, J., Shahmardani, M. & Rezazadeh, G. (2012). Free vibration of membrane/bounded incompressible fluid. *Applied Mathematics and Mechanics*, *33*(9), 1167-1178.

[62] Gorman, D., Horacek, J., Mulholland, A. & Gorman, M. N. (2013). Analysis and interpretation of compressible fluid interaction upon the vibration of a circular membrane. *Acta Technica, 58*(4), 443-461.

[63] Kolaei, A. & Rakheja, S. (2019). Free vibration analysis of coupled sloshing-flexible membrane system in a liquid container. *Journal of Vibration and Control, 25*(1), 84-97.

[64] Sygulski, R. (2007). Stability of membrane in low subsonic flow. *International Journal of Non-Linear Mechanics, 42*(1), 196-202.

[65] Sygulski, R. (1997). Numerical analysis of membrane stability in air flow. *Journal of Sound and Vibration, 201*(3), 281-292.

[66] Knight, J., Lucey, A. & Shaw, C. (2010). Fluid–structure interaction of a two-dimensional membrane in a flow with a pressure gradient with application to convertible car roofs. *Journal of Wind Engineering and Industrial Aerodynamics, 98*(2), 65-72.

[67] Wood, J. N., Breuer, M. & De Nayer, G. (2018). Experimental studies on the instantaneous fluid–structure interaction of an air-inflated flexible membrane in turbulent flow. *Journal of Fluids and Structures, 80*, 405-440.

[68] Ohyama, T., Tanaka, M., Kiyokawa, T., Uda, T. & Murai, Y. (1989). Transmission and reflection characteristics of waves over a submerged flexible mound. *Coastal Engineering in Japan, 32*(1), 53-68.

[69] Rojratsirikul, P., Wang, Z. & Gursul, I. (2009). Unsteady fluid–structure interactions of membrane airfoils at low Reynolds numbers. *Experiments in Fluids, 46*(5), 859.

[70] Liang, S. J., Neitzel, G. & Aidun, C. (1997). Finite element computations for unsteady fluid and elastic membrane interaction problems. *International Journal for Numerical Methods in Fluids, 24*(11), 1091-1110.

[71] Wood, C., Gil, A., Hassan, O. & Bonet, J. (2008). A partitioned coupling approach for dynamic fluid–structure interaction with applications to biological membranes. *International Journal for Numerical Methods in Fluids, 57*(5), 555-581.

[72] Le, D. V., White, J., Peraire, J., Lim, K. M. & Khoo, B. (2009). An implicit immersed boundary method for three-dimensional fluid–

membrane interactions. *Journal of Computational Physics*, *228*(22), 8427-8445.

[73] Quaini, A. & Quarteroni, A. (2007). A semi-implicit approach for fluid-structure interaction based on an algebraic fractional step method. *Mathematical Models and Methods in Applied Sciences*, *17*(06), 957-983.

[74] Sauer, R. A. & Luginsland, T. (2018). A monolithic fluid–structure interaction formulation for solid and liquid membranes including free-surface contact. *Computer Methods in Applied Mechanics and Engineering.*, *341*, 1-31.

In: Membrane Potential: An Overview   ISBN: 978-1-53616-743-6
Editor: Milan Marušić   © 2019 Nova Science Publishers, Inc.

Chapter 3

ESTIMATIONS AND ACTUAL MEASUREMENTS OF THE PLASMA MEMBRANE ELECTRIC POTENTIAL DIFFERENCE IN YEAST

Antonio Peña*, Norma Silvia Sánchez
and Martha Calahorra
Departamento de Genética Molecular,
Instituto de Fisiología Celular,
Universidad Nacional Autónoma de México,
Ciudad de México, México

ABSTRACT

Estimation of the plasma membrane potential (PMP) in yeast by the fluorescence changes of various indicators have been studied by several groups for many years; with most variable results. The most used of these

---

*Corresponding Author's E-mail: apd@ifc.unam.mx.

indicators have been $DiSC_3(3)$ and $DiSC_3(5)$. This contribution explores different dyes studied, methods and incubation media, as well as the different parameters analyzed. Particularly with $DiSC_3(3)$, from rather high to very low values of PMP have been calculated. Recently, we reported that fluorescence changes and accumulation of acridine yellow can be used to estimate and obtain actual values of the PMP in yeast. The experiments were performed with an old dye from a flask labeled "acridine yellow" from a commercial source. However, NMR and mass spectrometry revealed that it was thioflavin T. With the pure dye, the experiments were repeated. Also the accumulation of the dye was measured to obtain real values of PMP, mainly based in permeabilizing the cells with chitosan in the absence or presence of an adequate concentration of KCl that allowed to correct the raw data obtained. Hence, more accurate values were obtained. Moreover, results of comparing this dye with others used so far, point to thioflavin T as the best one to follow by fluorescence and measure by its accumulation the plasma membrane potential in yeast.

Keywords: yeast, plasma membrane potential, fluorescence, *S. cerevisiae,* thioflavin T

INTRODUCTION

The mechanism of K^+ transport in yeast was proposed as driven by the plasma membrane electric potential difference (PMP) generated by a H^+-ATPase (Peña et al. 1972; Peña A. 1975). This mechanism was also found in *Neurospora crassa* (Slayman, Long, and Lu 1973), later in many other yeasts and fungi (Goffeau and Slayman 1981), and also in plants (Sze 1983). Studies were also performed to measure the relationship between the changes produced by the ATPase on the external and internal pH when K^+ was added or the pH of the medium was increased (Peña A. 1975). The mechanism proposes that the H^+-ATPase produces a ΔpH, that has been measured (Peña et al. 1972), but also a plasma membrane potential difference (PMP), measured originally by the accumulation of tetraphenylphosphonium (Borst-Pauwels 1981; Vacata et al. 1981; De la Peña et al. 1982). Since then, other studies have been performed using the accumulation or the fluorescence changes of different dyes (De la Peña et al. 1982; Peña et al. 2010; Plášek et al. 2012; Plášek and Gášková 2014), but most of those results appear to be inaccurate.

More recently (Calahorra et al. 2017a; Peña et al. 2017), the use of acridine yellow to estimate by fluorescence the PMP changes in yeast, as well as the actual values by its accumulation, were reported. As mentioned, the dye was really thioflavin T (Calahorra et al. 2019). Thioflavin T was then purchased pure, and with small variations, also similar results were obtained.

In general, the mentioned reports of the actual PMP values obtained from the accumulation of the dyes or other molecules can be taken as inaccurate mostly because none of them considered that the concentration reached inside was not only due to the influence of the PMP, but also to their internal binding because of its hydrophobic and cationic nature. Previously (Calahorra et al. 2017a), we tried to define a numeric value for the yeast PMP by using supposedly acridine yellow. In those studies it was particularly difficult to obtain satisfactory values after the addition of K^+ because only the internal binding of the dye because of its hydrophobicity was considered, but not accurately that due to its cationic characteristics. Using the pure thioflavin T (Figure 1), previous results were confirmed, and using a larger concentration, as well as different conditions, more accurate and reliable results were obtained.

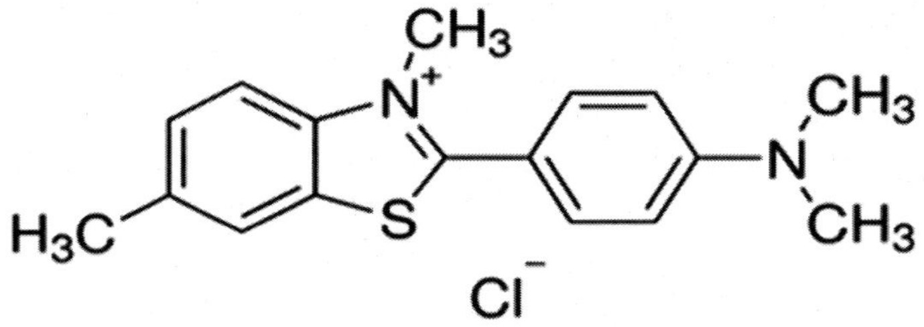

Figure 1. Chemical structure of thioflavin T.

From these results our proposal is now corrected, confirming that thioflavin T can be used to qualitatively follow and to obtain actual values and changes of the PMP in *S. cerevisiae*. Since those values could be obtained with a dye which among its properties is characterized by a lower hydrophobicity than others, experiments performed with other dyes are also presented and analyzed.

GENERAL PROCEDURES

As in any other procedure, the particular experimental conditions are central to the results obtained. Because of this, different factors involved are described.

Strain and Growth Conditions

Most reports are referred to experiments performed with *Saccharomyces cerevisiae*, usually grown in YPD and collected in the exponential phase. In our experiments, this yeast from a commercial strain (La Azteca, México) obtained form an isolated single colony has always been used. Cultures are started by placing a loophole of cells in 500 mL of liquid YPD grown for 24 h at constant room temperature (30°C) in an orbital shaker at 250 rpm. In our experiments, in order to be able to see the effects of a substrate (glucose) to energize the cells, they are starved by collecting them by centrifugation, suspending them in 250 mL of water and incubating in the same shaker for 24 h. After this incubation the cells are collected by centrifugation, washed once with water and suspended in water at a ratio of 0.5 g (wet weight)$\cdot$mL^{-1}.

Dyes Used

The dyes more frequently used have been DiSC$_3$(3) (3,3'-Dipropyl thiacarbocyanine iodide), DiSC$_3$(5) (3,3'-Dipropylthiadicarbocyanine iodide), DiOC$_6$(3) (3,3'-Dihexyloxacarbocyanine iodide) (Sigma), but also acridine yellow (Sigma), acridine orange, safranine O, neutral red, rhodamine G and thioflavin T (Biotium) have been tested under the conditions described for each group of results.

METHODS

The procedures used have been: a) following the fluorescence changes of the dyes; b) analyzing their distribution within the cell, and c) measuring the accumulation of the dyes by the cells, but also using some correction factors due to the fact that they are not only accumulated due to the PMP, but they also bind to the internal components of the cells.

Membrane Potential Estimations by Fluorescence

As done by many groups, membrane potential changes have been estimated by measuring the fluorescence changes of the different dyes at their maximum excitation-emission wavelengths. Some authors (Plášek et al. 2012; Plášek and Gášková 2014) have actually used the changes of the absorption maxima to estimate the values, but also using complex incubation media, and have reported rather low values. We have used a more simple approach, by following the fluorescence intensity changes under different conditions and after the addition of agents known to affect both the distribution and accumulation of the dyes inside the cells, mainly in the mitochondria, cytoplasm and vacuole. We believe that the incubation medium has to be as simple as possible, and free of components that may affect the PMP. It is 10 mM MES-TEA buffer, pH 6.0; 20 mM glucose, and 10 μM $BaCl_2$ to avoid the binding to the surface of the cells; final volume, 2.0 mL. Besides, changes have been estimated by adding 5 μL 3% H_2O_2, 10μM CCCP (Carbonylcyanide-*m*-phenylhydrazine), or 10 mM KCl. As mentioned, fluorescence is followed at the corresponding maxima of excitation and emission wavelengths, using an SLM Aminco spectrofluorometer with stirring and temperature regulation at 30°C. The composition of the incubation medium is an important factor in the collection of the spectral changes (Calahorra et al. 2017b; Peña et al. 2010).

Microscopic Fluorescence Images

Fluorescence images were obtained by microscopy, providing more information about the localization of the dyes under different conditions. Microscope fluorescence images have been obtained, both with $DiSC_3(3)$ (Peña et al. 2010) and with pure thioflavin T obtained from Biotium. This combined approach allows to define what is supposed to produce the fluorescence changes.

Accumulation of Dyes

To obtain the actual PMP values, the accumulation of thioflavin T or other dyes during 10 min was measured under different conditions. In all cases, the essential medium was the same already mentioned, and the indicated concentrations of the dyes, in a final volume, 3.0 mL. Where indicated, 10 μM CCCP and, or 100 μg chitosan, low molecular weight, without or with 200 mM KCl was added. The use of an adequate incubation medium, particularly the buffer has been discussed before (Peña et al. 2010).

After equilibrating the tubes to 30°C in a water bath, 150 mg of cells were added, a) with glucose; b) with glucose plus 10 μM CCCP; c) with 10 mM KCl, and the cells were centrifuged after 10 min. In order to correct the accumulation values for the total binding of the dye to the internal components of the cells, the amount of dye remaining inside was measured as described in (b), but adding 100 μg of chitosan, which allows (by permeabilizing the cells) to quantify that total dye bound inside due to its cationic and hydrophobic nature. To obtain the value of the binding because of the cationic nature of the dye, cells were also incubated as described in (b), but in the presence of 100 μg of chitosan and 200 mM KCl, assuming that in this way the bound dye can be displaced from probable anionic sites; the remaining dye after KCl addition was taken as that bound only to hydrophobic sites. Chitosan used was low molecular weight (Sigma), dissolved in water to a concentration of 10 mg.mL^{-1} by the addition of HCl to a pH around 4.8.

In all cases, using a DW2a SLM Aminco-Olis spectrophotometer, the absorbance values at their maximum absorption lambda of adequate dilutions were recorded and concentration was calculated by comparing with the absorbance of a standard curve of each.

K⁺ Efflux and Uptake

It was found that several of the cationic dyes affect the response of a K^+ selective electrode, so the efflux they produced was followed by incubating cells as follows. Cells (150 mg wet weight) were incubated for 10 min at 30°C in 10 μM MES-TEA, pH 6.0, 20 μM glucose with 200 μM of each dye, in a final volume of 5 mL. After incubation, cells were centrifuged and K^+ in the supernatant was measured by flame photometry (Carl Zeiss PF5). Total K^+ in the cells was obtained by boiling an equal number of cells with no dye added, during 10 min.

To measure the uptake, after incubating in a similar medium, but with 5 mM KCl, the tubes were cooled in ice, centrifuged, and washed with 4.0 mL of ice cold water also by centrifugation. The final pellet obtained was resuspended in 5.0 mL of water and placed in boiling water for 10 min. After centrifugation, the K^+ concentration was measured in the supernatant. To calculate the internal concentration, an internal water volume of 0.45 mL.g⁻¹ of cell wet weight was considered (Sánchez et al. 2008).

Calculation of the Vacuolar and Cytoplasmic Volumes

In previous and recent experiments microscopic images showed that thioflavin T, under basal conditions (only with glucose), does not accumulate in the vacuole, so it was required to calculate the vacuolar and cytoplasmic volumes as described before. Shortly, from images obtained in the microscope, the mean relative radii of vacuoles and cytoplasm for 30 cells were obtained with Image J software. Knowing the internal water volume of the cells (Sánchez et al. 2008), the values of the internal water volumes could be obtained, with the results shown in Table 1.

Table 1. Calculation of the internal volumes

Total cell water per g of cells	0.372 mL (Sánchez et al. 2008)
Mean radii of 30 cell images	0.821 relative value
Mean radii of their vacuole images	0.404 relative value
Apparent mean cell volume	2.318
Apparent mean vacuole volume	0.276
Vacuole/cell volume ratio	0.119
Total vacuole water per g cells	0.0442 mL
Cytoplasm volume per g of cells	0.3277 mL
Total water for 150 mg cells	0.0558 mL
Cytoplasm volume for 150 mg cells	0.0492 mL
Vacuole volume for 150 mg cells	0.00663 mL
Cytoplasm/total volume ratio	0.882

RESULTS

Fluorescence Changes

Changes obtained with pure thioflavin T are shown in Figure 2, when added to yeast cells at similar concentrations (15 μM) to those used before (Calahorra et al. 2017a).

The result was similar to that reported before with $DiSC_3(3)$ (Peña et al. 2010), and also with the supposed acridine yellow (Calahorra et al. 2017a; Calahorra et al. 2019), now thioflavin T. A low increase of fluorescence that reached a constant value after approximately 2 min, that can be interpreted as a result of the accumulation of the dye by the mitochondria, where due to the high concentration reached, its fluorescence is quenched. It was then followed by a small increase when oxygen was exhausted, due to a partial deenergization of the mitochondria, that however still can maintain its membrane potential by using the ATP synthesized by glycolysis. In fact, this small increase can be reversed by adding hydrogen peroxide. Then a small concentration of CCCP (10 μM) added to fully deenergize the mitochondria, produced a large fluorescence increase, that was followed by a large and slow decrease of fluorescence after adding 20 mM KCl.

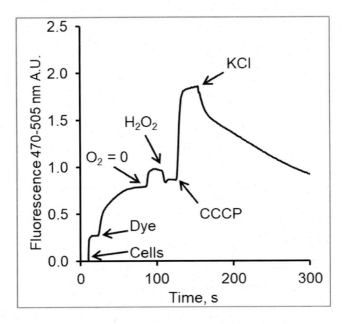

Figure 2. Fluorescence changes of thioflavin T with yeast cells. The incubation medium was 10 mM MES-TEA buffer, pH 6.0, 20 mM glucose, 10 μM BaCl$_2$, to a final volume of 2.0 mL. Tracings show the changes after the addition of 50 mg of yeast cells; 15 μM thioflavin T; the exhaustion of oxygen (O$_2$= 0); the addition of 5μL of 3% H$_2$O$_2$; the addition of 10 μM CCCP, and the addition of 20 mM KCl. Fluorescence changes were followed in an SLM spectrofluorometer at 470 nm excitation and 505 nm emission wavelengths respectively.

Microscopic Images

Consistent with these results, the images in Figure 3 show: a) with glucose alone, a rather low fluorescence is located in the mitochondria, in agreement with the idea that these organelles highly concentrate the dye to a degree that results in quenching of fluorescence, which also happens with DiSC$_3$(3) (Peña et al. 2010); b) after the addition of 10 μM CCCP, there is a large increase of fluorescence, no longer concentrated in the mitochondria, consistent with the large fluorescence observed after the addition of the uncoupler in the spectrofluorometer tracings; c) the further addition of 20 mM KCl produced a large decrease of the observed fluorescence.

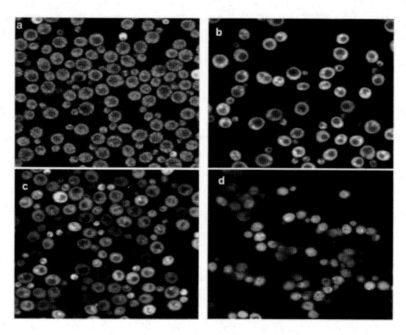

Figure 3. Fluorescence images of cells observed by microscopy under different conditions. Images were obtained as described under methods using the same medium described for Figure 2, containing 10 mM MES-TEA buffer, pH 6.0, 10 μM BaCl$_2$, 20 mM glucose, 50 mg of cells, wet weight, followed by the dye (15 μM). a: image obtained approximately 5 min after adding the cells in the buffered medium with BaCl$_2$ and glucose. b: image from the same cells approximately 5 min after adding 10 μM CCCP. c: image obtained 5 min after the addition of 20 mM KCl and d: image obtained 5 min after the addition of 100 μg chitosan and 200 mM KCl.

After the addition of chitosan (100 μg), also a decrease, but not the disappearance of fluorescence was observed, indicating that still a significant amount of the dye remains bound inside the cell, due to its hydrophobic and cationic nature. The combined addition of chitosan and 200 mM KCl produced a further decrease of fluorescence (data not shown), but not its disappearance, indicating that under these conditions, although this K$^+$ concentration displaces the dye bound due to its cationic nature, part of it still remains because of its hydrophobic character. In the absence of chitosan, the dye did not enter the vacuole, but in its presence, it distributed uniformly inside the cell d). Results are perfectly consistent with those presented before (Calahorra et al. 2017a), although a lower concentration was used.

Other Dyes

Also other dyes have been used to follow the fluorescence changes; most of those tested show similar changes to those described for thioflavin T. Tracings are shown in Figure 4.

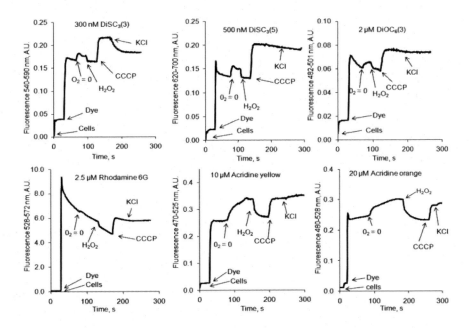

Figure 4. Fluorescence changes of several dyes upon their interaction with yeast cells. The incubation conditions were the same used in Figure 2. Fluorescence was followed at the maximal excitation-emission wavelengths of the respective dyes. The concentration used is indicated in each case, obtained as the best by following the changes with that concentration producing the clearest changes. Similar changes were observed in the other figures, except for the changes produced after the addition of KCl.

In all cases, most changes were similar to those observed with thioflavin T, however, probably the clearest difference was that the response to the addition of KCl was not observed, except in the case of $DiSC_3(3)$. Another interesting finding was that with rhodamine 6G, all concentrations tested showed a decrease in its fluorescence, probably because this dye requires higher concentrations to observe the changes that concentrate it in the cell, with the consequent quenching of fluorescence.

PMP VALUES OBTAINED BY
THE ACCUMULATION OF DYES

Corrections

Previous PMP values (in millivolts) obtained by measuring the accumulation of $DiSC_3(3)$ at two concentrations in the presence of 10 µM CCCP (Peña et al. 2010) were around -168 mV and -140 mV when 5 mM KCl was added

The measurement of the dye accumulation at concentrations similar to those of the fluorescence experiments produces uncertain results because practically all the dye is taken, as shown by the already described images of the cells. In new experiments, using thioflavin T, a larger concentration was used, with more reliable results. Since the microscope images (Figure 3) and also previous accumulation experiments, clearly showed that mitochondria concentrate the dye, this was avoided by the addition of 10 µM CCCP. In the first experiment performed as described, but using 300 µM thioflavin T, apparent values for the PMP of -220 mV with glucose were obtained, which decreased to -184.7 mV by the addition of CCCP, and to -146.8 mV after the further addition of 10 mM KCl. It is so clear that values with glucose alone are overestimated because a large part of the dye accumulates in the mitochondria, but also that even in the presence of CCCP, the amount of dye inside the cells is not only due to the PMP difference, but also to its binding to the internal components of the cells.

To correct the values obtained in the presence of glucose and CCCP, adding chitosan, that permeabilizes the cells (Peña et al. 2013), a large amount of the dye that remains bound to their internal components can be estimated, and this is then subtracted from the total accumulated in the absence of chitosan.

In previous work (Calahorra et al. 2017a), in cells permeabilized with chitosan, the amount of dye bound because of its cationic nature was estimated by attempting to displace it by the addition of 10 mM KCl, but this concentration is not enough to displace all the dye bound to anionic sites. In intact cells, after the addition of 10 mM KCl, the internal K^+ concentration reaches around 200 mM; for that reason, in the new experiments, after the addition of chitosan, where the cells are already permeabilized, that concentration of KCl (200 mM) was added to achieve maximal displacement of the dye.

In the Presence of Glucose

To correct the values obtained for the apparent PMP with glucose plus, the cytoplasmic bound dye in the cells was measured after the addition of 100 µg of chitosan, and this concentration was then subtracted from the total accumulated in the cytoplasm of intact cells in the presence of glucose (plus CCCP) to calculate what was considered the actual PMP value. The corrected internal concentration divided by the external one obtained for glucose plus CCCP gave an internal/external concentration ratio whose negative log is -2.67, which multiplied by 60 results in a corrected value of -173.0 mV.

Glucose + KCl

The corrected PMP values obtained when chitosan and 200 mM KCl were added were obtained as follows: The external concentration obtained after chitosan +200 mM KCl allowed estimating the amount of dye remaining within the cell due to its hydrophobicity, and by the difference from the total amount in the absence of KCl, that bound because of its cationic nature. The corrected internal/external concentration ratio in the presence of glucose plus KCl in intact cells was obtained, and from its negative log, a PMP value of -117.8 mV was calculated. Only values in the presence of CCCP are presented because in its absence they are influenced by the large accumulation by the mitochondria.

Table 2. Values of the PMP (in mV) of yeast cells under different conditions. Raw and corrected values. n=5

Condition	Raw values	Corrected values
Glucose-CCCP	-184.7 ± 3.6	-173.0 ± 9.6
Glucose-CCCP-KCl	-146.8 ± 3.7	-117.8 ± 3.5

The Use of Other Dyes

One of the main differences of thioflavin T and other dyes is that its hydrophobicity, measured as the distribution coefficient between water and dichloromethane is as follows:

Table 3. Distribution coefficients of several dyes between dichloromethane and buffer

Dye	Distribution Coefficient
Thioflavin T	9.4
DiSC$_3$(3)	26.6
DiSC$_3$(5)	893.4
DiOC$_6$(3)	164.3
Safranin O	0.36
Neutral Red	3.37
Rhodamine 6 G	2.16
Acridine yellow	12.5
Acridine orange	0.94

Hydrophobicity of dyes measured as their distribution coefficients between 2 mL of 10 mM MES-TEA, pH 6, and 2 mL of dichloromethane. The dyes, 120 μM, were added and vigorously stirred in a vortex mixer. Then the mixture was centrifuged for 5 min at 3000 RPM. The absorbance of the water layer was measured at the maximal lambda value of each dye and its concentration calculated by comparison with a standard curve.

Clearly, thioflavin T has a low distribution coefficient between dichloromethane and water, indicating that it is among the least hydrophobic of the dyes tested, and values of the PMP could be obtained from its accumulation. $DiSC_3(3)$, with the lowest hydrophobicity of other similar dyes and higher than those of thioflavin T. That dye, as shown in Table 4 was tested, and gave results not very different from those with thioflavin T. Results were then compared with other dyes at 200 µM: rhodamine 6G, safranin O, neutral red, real acridine yellow, and acridine orange.

**Table 4. Raw and corrected values of the PMP
of yeast cells with different dyes**

	Raw values	Corrected values
Thioflavin T 333 µM (n = 5)		
Glucose-CCCP	-184.7 ± 3.6	-173.0 ± 9.6
Glucose-CCCP-KCl	-146.8 ± 3.7	-117.8 ± 3.5
$DiSC_3(3)$ 167 µM (n = 3)		
Glucose-CCCP	-210.4 ±3.9	-181.9 ± 5.9
Glucose -CCCP-KCl	-202.0 ±4.9	-166.5 ± 7.8
Rhodamine 6G 200 µM (n = 3)		
Glucose - CCCP	-194.9 ± 9.8	-173.3 ± 9.5
Glucose -CCCP-KCl	-147.1 ± 17.4	-155.2 ± 16.7
Safranin 200 µM (n = 3)		
Glucose-CCCP	-169.1 ± 1.8	-146.1 ± 2.1
Glucose-CCCP-KCl	-150.0 ± 3.2	-163.0 ± 2.9
Neutral red 200 µM (n = 3)		
Glucose-CCCP	-173.3 ± 25.3	-149.7 ± 36.2
Glucose -CCCP -KCl	-138.2 ± 31.4	-154.7 ± 31.6
Acridine yellow 300 µM (n = 4)		
Glucose -CCCP	-209.8± 8.5	-195.4 ± 9.6
Glucose CCCP KCl	-155.4± 12.0	-141.9 ± 9.7

Values were obtained as described in the text for thioflavin T at the indicated concentrations. Means of 3 to 5 experiments, each ± standard deviation.

Most of them gave rather similar results, except for safranin O and neutral red, with values that after correction by the addition of KCl, in the presence of chitosan, for some reason we cannot explain, were higher than in its absence. Results with acridine orange are not presented because it gave similar results under all the different conditions.

Collateral Effects of High Concentrations of Thioflavin T on the Acidification Capacity, K⁺ Transport, and Respiration

Particularly in the case of thioflavin T, concentrations used for the accumulation experiments (333 μM), higher than those used before, require observing their effects on three important physiological parameters: the capacity of the cells to acidify the medium, to take up potassium, and respiration. In the case of thioflavin T, it did not inhibit acidification of the medium, but on the contrary, increased it for about two minutes, and then returned to a rate almost similar to that of the control. One of the main *S. cerevisiae* functions, proton pumping, on which many other transport systems depend, was not inhibited, but also increased, probably because of the energy required for the uptake of the dye at these concentrations. Regarding K⁺ uptake, tracings could not be obtained because the K⁺ selective electrode responds to the addition of the dye. However, the effect of all the dyes tested was also estimated on two parameters: a) the uptake of K⁺, adding 5 mM KCl to the incubation media, followed by measuring its accumulation in the presence of 200 μM of each dye, and b) the efflux of K⁺ produced under the same conditions, but in the absence of externally added KCl. Results are shown in Figure 5.

Some dyes produced a small increase of the K⁺ efflux, and mainly acridine yellow and acridine orange clearly increased the efflux of the monovalent cation. This corresponded in a reverse way with the uptake of K⁺.

It was also found that thioflavin T did not inhibit respiration, but stimulated it, and CCCP still further stimulated it (not shown). This means that at these concentrations (100 μM to 300 μM) the presence of the dye results in an increase of both basal and maximal (uncoupled) mitochondrial respiratory capacity.

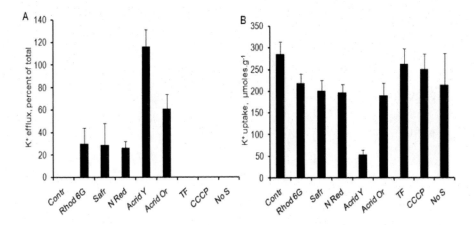

Figure 5. Effects of different dyes at a 200 µM concentration on K⁺ uptake and efflux. (A) Efflux in a medium without KCl, and (B) Uptake in the presence of 5 mM K⁺, as described in Methods. Contr = Control, Rhod 6G = Rhodamine 6G, Safr = Safranine, N Red = Nile red, Acrid Y = Acridine yellow, Acrid Or = Acridine orange, TF = Thioflavine, No S = no substrate added. Values are means ± standard deviation of 3 to 4 different experiments.

Particular Comments on Dyes

Although it would appear that any fluorescent or colored dye might be used to actually measure the PMP of yeast cells with the adequate corrections described, there are some pertinent observations on several of them: $DiSC_3(3)$, $DiSC_3(5)$ and $DiOC_6(3)$. Although the first is the least hydrophobic of those tested, in experiments performed to correct the values for its displacement by KCl it was found that after the treatment with chitosan, a large part of it remains bound to the cells, and only a very small amount could be displaced by the high concentrations of KCl. This should be a valuable argument to eliminate it, as well as $DiSC_3(5)$ and $DiOC_6(3)$, which are even more hydrophobic as useful for the quantitative measurement of the PMP.

Safranin O and neutral red. Because of the unexplainable behavior, with the corrected values higher in the presence of KCl than in its absence, they are not to be adequate for the purpose of measuring the real values of the PMP in yeast.

Rhodamine 6G. This dye appeared to be also adequate, with similar values to those for thioflavin T, except for the fact that fluorescence intensity did not decrease upon the addition of KCl.

Acridine yellow. Although values obtained are more or less close to those with thioflavin T, the corrected figures for the PMP in the absence of KCl appear to be rather high.

Thioflavin T. This is a cationic dye with a low hydrophobicity, and what appears to be a balance between these two properties, that may be proposed to measure the actual values of the PMP in yeast, given the correction factors used.

CONCLUSION

It would appear that in principle, any more or less hydrophobic cationic dye could be used to in fact measure the real PMP difference in yeast; results point to thioflavin T as the best dye for that purpose. It should be made clear that the dye used in the previously reported experiments (Calahorra et al. 2017a, b; Peña et al. 2017) was not acridine yellow, but thioflavin T (Calahorra et al. 2019).

The use of thioflavin T allowed the best approximation to the estimation by fluorescence and measurement by its accumulation, of the PMP in *S. cerevisiae*. Results also show that neither the fluorescence changes nor the simple accumulation values under different conditions can be taken as such without considering the uptake of the dye by the mitochondria and its binding due to both its hydrophobic and cationic nature. Already the fluorescence changes observed indicate that a large proportion of the dye was accumulated by the mitochondria, and a low concentration of CCCP (10 to 20 µM) is enough to deenergize the mitochondria without an effect on the plasma membrane of yeast (Peña 1975). This fortunate situation allows eliminating the participation of the mitochondria. This conception was supported by the observation of the fluorescence images that clearly showed that with glucose alone, although the cells captured a large proportion of the dye, it was located largely in the mitochondria, and fluorescence was faint because the high accumulation resulted in quenching. Also from the microscope images it could be concluded that after permeabilizing the cells with chitosan (not shown) an important portion of the dye remained inside imputable to its hydrophobic and cationic characteristics.

From these arguments, it was obvious that by permeabilizing the cells with chitosan, the total amount of bound dye within the cells could be measured. But not only this, also in the cells permeabilized with chitosan, adding a large enough concentration of KCl (200 mM), the dye bound to anionic sites could be displaced, remaining only that bound to hydrophobic molecules; then, by difference the cationic binding could be obtained. In this way the concentration values obtained with glucose plus CCCP could be corrected by subtracting from the total accumulated. In the case of the cells incubated with 10 mM KCl the situation is the reverse: the uptake of K^+ results in an internal dye concentration decrease not only owed to the decrease of the PMP by the entrance of positive charges, but also because it results in an accumulation of around or more than 200 mM of the cation, which displaces the dye from its binding to anionic sites. By the procedure described, the amount and concentration of the dye displaced by K^+ when in intact cells 10 mM KCl was added could be obtained; then this concentration of the dye could be added to that found in intact cells incubated in the presence of KCl (with CCCP). These correction factors can be obtained with enough approximation and used to calculate the real PMP values. It is important to remark that all values of the PMP require the presence of 10 to 20 μM CCCP to eliminate the additional and very large accumulation of the dye by the mitochondria.

The results obtained are higher than those reported by other authors (Vacata et al. 1981); (De la Peña et al. 1982), and definitely much higher than those proposed by (Plášek et al. 2012) and (Plášek and Gášková 2014). The two latter reports also have a problem, they assume that the PMP of yeast is collapsed by a low concentration (10 μM) of either CCCP or FCCP, which as mentioned is not enough to increase the H^+ conductivity of the plasma membrane of yeast. This also can be confirmed by the experiments shown in Figure 5, in which it is clear that CCCP at the concentration used (10 μM) did not affect the uptake, neither increased the K^+ efflux, that would be expected if the uncoupler at that concentration affected the PMP.

With the corrected values, values were lower in the presence of KCl (Table 2), but the decrease was perhaps not so large as expected from the fluorescence changes. This can be explained because although in the presence of K^+, a large decrease of PMP would be expected, former studies (Peña et al. 1969; Peña 1975) showed that the addition of the cation immediately results in an increase of the H^+ pumping by the H^+-ATPase of the plasma membrane. Then this expenditure of ATP produces an increase of ADP which is responsible for the acceleration of glycolysis (and respiration), and by that the ATP levels are recovered. It is so that the H^+-ATPase activity, supported by an increased rate of glycolysis can maintain the PMP values not much lower than those present in the absence of K^+. In the mentioned studies it was found that in fact, glycolysis is accelerated in the presence of potassium, but also that the acceleration persists after K^+ has been taken up by the cells, probably because energy is necessary not only to capture the cation; after this happens, a new equilibrium is established between its uptake and efflux, due to the high concentrations reached inside. In fact, Rothstein and Bruce (1958) described that after K^+ has been taken up by yeast a steady state is established in which an efflux-influx equilibrium takes place, that not surprisingly, is accompanied by the partial recovery of the PMP at the expense of an accelerated glycolysis. It is so that the H^+-ATPase activity, supported by an increased rate of glycolysis can maintain the PMP values not much lower than those present in the absence of K^+.

Using a similar procedure; i.e., with the correction factors, $DiSC_3(3)$, the least hydrophobic of the cyanines, produced results not far different from those with thioflavin T, even though it binds inside with a large affinity because of its hydrophobicity. On the other hand, using 200 μM safranin O, rhodamine 6G and neutral red, for some reason, values obtained in the presence of CCCP were lower than those obtained with thioflavin T. Also, we cannot explain why values after adding KCl were higher than in its absence.

Another dye used before is ethidium bromide, but it has a problem, it appears to be transported by the same carrier as K^+ (Peña and Ramírez 1975). It seems that even small variations in structure make some dyes unsuitable, even for the qualitative estimation of the PMP.

All previous results reported regarding the quantitative value of the PMP in yeast cells are probably wrong because they did not consider the internal binding of the agents used under different conditions.

In summary, the studies performed allow to propose the use of thioflavin T to estimate by fluorescence, and measure by its corrected accumulation, the actual PMP values of yeast cells.

ACKNOWLEDGMENTS

This work was supported by grants 238497 from the Consejo Nacional de Ciencia y Tecnología, México, and grants IN223999, and IN202103 from the Universidad Nacional Autónoma de México. We thank the assistance of Dr. Roberto Arreguín and María Vanegas, from the Chemistry Institute of our University, through whom we received the assistance of Dr. Nuria Esturau from the NMR Laboratory of the Institute of Chemistry of our University, as well as that of Dr. Benjamín Velasco, from the Laboratorio Nacional de Prevención y Control de Dopaje, part of the Comisión Nacional de Cultura Física y Deporte, México; these two laboratories succeeded in defining the real structure of the dye used. The authors also thank Dr. Yazmín Ramiro Cortés of our Institute for her invaluable help obtaining the microscope images.

CONFLICT OF INTEREST

The authors have declared that no conflict of interest exists.

REFERENCES

Borst-Pauwels, G.W.F.H. (1981). "Ion Transport in Yeast." *Biochimica et Biophysica Acta (BBA)-Reviews on Biomembranes* 650 (2–3). Elsevier: 88–127. doi:10.1016/0304-4157(81)90002-2.

Calahorra, M., Sánchez N.S., and Peña A. (2017a). "Acridine Yellow. A Novel Use to Estimate and Measure the Plasma Membrane Potential in *Saccharomyces cerevisiae.*" *Journal of Bioenergetics and Biomembranes* 49 (3). doi:10.1007/s10863-017-9699-7.

Calahorra, M., Sánchez N.S., and Peña A. (2017b). "Effects of acridine derivatives on Ca^{2+} uptake by *Candida albicans*." *Bioenergetics Open Access* 6 (2): 151. doi:10.4172/2167-7662.1000151.

Calahorra, M., Sánchez N.S., and Peña A. (2019). "Retraction note to: Acridine Yellow. A Novel Use to Estimate and Measure the Plasma Membrane Potential in *Saccharomyces cerevisiae.*" *Journal of Bioenergetics and Biomembranes* May 31. doi:10.1007/s10863-019-09801-y.

De la Peña, P., Barros F., Gascón S., Ramos S. and Lazo P.S. (1982). "The Electrochemical Proton Gradient of *Saccharomyces.*" *European Journal of Biochemistry* 123 (2): 447–53. doi:10.1111/j.1432-1033.1982.tb19788.x.

Goffeau, A., and Slayman C.W. (1981). "The Proton-Translocating ATPase of the Fungal Plasma Membrane." *Biochimica et Biophysica Acta (BBA) - Reviews on Bioenergetics* 639 (3–4). Elsevier: 197–223. doi:10.1016/0304-4173(81)90010-0.

Peña, A., Cinco G., Gómez Puyou A., and Tuena M. (1969). "Studies on the Mechanism of the Stimulation of Glycolysis and Respiration by K^+ in *Saccharomyces cerevisiae.*" *BBA-Bioenergetics.* doi:10.1016/0005-2728(69)90187-X.

Peña, A., Cinco G., Gómez Puyou A., and Tuena M. (1972). "Effect of the pH of the Incubation Medium on *S. cerevisiae.*" *Archives of Biochemistry and Biophysics* 153 (4): 413–25. doi:10.1016/0003-9861(72)90359-1.

Peña, A, and Ramírez G. (1975). "Interaction of Ethidium Bromide with the Transport System for Monovalent Cations in Yeast." *The Journal of Membrane Biology* 22 (1): 369–84. doi:10.1007/BF01868181.

Peña, A. (1975). "Studies on the Mechanism of K^+ Transport in Yeast." *Archives of Biochemistry and Biophysics* 167 (2). Academic Press: 397–409. doi:10.1016/0003-9861(75)90480-4.

Peña, A., Sánchez N.S., and Calahorra M. (2010). "Estimation of the Electric Plasma Membrane Potential Difference in Yeast with Fluorescent Dyes: Comparative Study of Methods." *Journal of Bioenergetics and Biomembranes* 42 (5): 419–32. doi:10.1007/s10863-010-9311-x.

Peña A, Sánchez NS, Calahorra M. (2013). "Effects of Chitosan on *Candida albicans*: Conditions for Its Antifungal Activity." *BioMed Research International* 2013. doi:10.1155/2013/527549.

Peña A, Sánchez NS, Calahorra M. (2017). "The Plasma Membrane Electric Potential in Yeast: Probes, Results, Problems, and Solutions: A New Application of an Old Dye?" *Old Yeasts - New Questions*. doi:10. 5772/intechopen.70403.

Plášek, J., Gášková D., Lichtenberg-Fraté H., Ludwig J., and Höfer M. (2012). "Monitoring of Real Changes of Plasma Membrane Potential by $DiSC_3(3)$ Fluorescence in Yeast Cell Suspensions." *Journal of Bioenergetics and Biomembranes* 44 (5): 559–69. doi:10.1007/s10863-012-9458-8.

Plášek, J. and Gášková D. (2014). "Complementary Methods of Processing $DiSC_3(3)$ Fluorescence Spectra Used for Monitoring the Plasma Membrane Potential of Yeast: Their Pros and Cons." *Journal of Fluorescence* 24 (2). Springer New York LLC: 541–47. doi:10. 1007/s10895-013-1323-6.

Rothstein A and Bruce M. (1958). "The potassium efflux and efflux in yeast at different potassium concentrations". *Journal of Cellular and Comparative Physiology* 51:145-159.

Sánchez, N.S., Arreguín R., Calahorra M., and Peña A. (2008). "Effects of Salts on Aerobic Metabolism of *Debaryomyces hansenii*." *FEMS Yeast Research* 8 (8): 1303–12. doi:10.1111/j.1567-1364.2008.00426.x.

Slayman, C.L., Long W.S., and Lu C.Y.H. (1973). "The Relationship between ATP and an Electrogenic Pump in the Plasma Membrane of *Neurospora crassa.*" *The Journal of Membrane Biology.* doi:10.1007/BF01868083.

Sze, H. (1983). "Proton-Pumping Adenosine Triphosphatase in Membrane Vesicles of Tobacco Callus: Sensitivity to Vanadate and K⁺." *Biochimica et Biophysica Acta (BBA) - Biomembranes* 732 (3). Elsevier: 586–94. doi:10.1016/0005-2736(83)90235-3.

Vacata, V., Kotyk A., and Sigler K. (1981). "Membrane Potential in Yeast Cells Measured by Direct and Indirect Methods." *Biochimica et Biophysica Acta (BBA) - Biomembranes* 643 (1): 265–68. doi: https://doi.org/10.1016/0005-2736(81)90241-8.

INDEX

CALMODULIN: STRUCTURE, MECHANISMS AND FUNCTIONS

EDITOR: Vahid Ohme

SERIES: Cell Biology Research Progress

BOOK DESCRIPTION: In *Calmodulin: Structure, Mechanisms and Functions*, the authors consider small and poorly-studied groups of plant calcium-dependent protein kinases that directly interact with calmodulin molecules.

SOFTCOVER ISBN: 978-1-53614-948-7
RETAIL PRICE: $82

FLAGELLA AND CILIA: TYPES, STRUCTURE AND FUNCTIONS

EDITOR: Rustem E. Uzbekov

SERIES: Cell Biology Research Progress

BOOK DESCRIPTION: Motility is an inherent property of living organisms, both unicellular and multicellular. One of the principal mechanisms of cell motility is the use of peculiar biological engines – flagella and cilia. These types of movers already appear in prokaryotic cells. However, despite the similar function, bacteria flagellum and eukaryote flagella have fundamentally different structures.

SOFTCOVER ISBN: 978-1-53614-333-1
RETAIL PRICE: $95

To see a complete list of Nova publications, please visit our website at www.novapublishers.com

Related Nova Publications

MITOGEN-ACTIVATED PROTEIN KINASES (MAPKs): ACTIVATION, FUNCTIONS AND REGULATION

EDITOR: Charles K. Hester

SERIES: Cell Biology Research Progress

BOOK DESCRIPTION: *Mitogen-Activated Protein Kinases (MAPKs): Activation, Functions and Regulation* opens with a summary of the present knowledge about MAPK, with special emphasis on p38 and c-Jun N–terminal kinase. The authors focus on how these signaling pathways are engaged during some infections with intracellular parasites.

SOFTCOVER ISBN: 978-1-53616-138-0
RETAIL PRICE: $69

BETA-GALACTOSIDASE: PROPERTIES, STRUCTURE AND FUNCTIONS

EDITOR: Eloy Kras

SERIES: Cell Biology Research Progress

BOOK DESCRIPTION: In *Beta-Galactosidase: Properties, Structure and Functions*, the authors discuss the main microorganisms that produce β-galactosidase, the characteristics of the culture media, bioprocessing parameters, the most relevant downstream steps used in the recovery of microbial β-galactosidase, as well as the main immobilization techniques.

SOFTCOVER ISBN: 978-1-53615-605-8
RETAIL PRICE: $95

To see a complete list of Nova publications, please visit our website at www.novapublishers.com